1994

Riemannian Geometry

Baum, R. J., *Philosophy and Mathematics*
ISBN 0-86720-514-2

Eisenbud, D., and Huneke, C., *Free Resolutions in Commutative
Algebra and Algebraic Geometry*
ISBN 0-86720-285-8

Epstein, D. B. A., *et al.*, *Word Processing in Groups*
ISBN 0-86720-241-6

Epstein, D. B. A., and Gunn, C., *Supplement to Not Knot*
ISBN 0-86720-297-1

Eves, H., *Fundamentals of Modern Elementary Geometry*
ISBN 0-86720-247-5

Geometry Center, University of Minnesota, *Not Knot* (VHS video)
ISBN 0-86720-240-8

Gleason, A., *Fundamentals of Abstract Analysis*
ISBN 0-86720-209-2

Hansen, V. L., *Geometry in Nature*
ISBN 0-86720-238-6

Harpaz, A., *Relativity Theory*: *Concepts and Basic Principles*
ISBN 0-86720-220-3

Loomis, L. H., and Sternberg, S., *Advanced Calculus*
ISBN 0-86720-122-2

Morgan, F., *Riemannian Geometry*: *A Beginner's Guide*
ISBN 0-86720-242-4

Protter, M. H., and Protter, P. E., *Calculus, Fourth Edition*
ISBN 0-86720-093-6

Redheffer, R., *Differential Equations*: *Theory and Applications*
ISBN 0-86720-200-9

Redheffer, R., *Introduction to Differential Equations*
ISBN 0-86720-289-0

Ruskai, M. B., *et al.*, *Wavelets and Their Applications*
ISBN 0-86720-225-4

Serre, J.-P., *Topics in Galois Theory*
ISBN 0-86720-210-6

RIEMANNIAN GEOMETRY
A BEGINNER'S GUIDE

Frank Morgan
Department of Mathematics
Williams College
Williamstown, Massachusetts

Illustrated by James F. Bredt
Department of Mechanical Engineering
Massachusetts Institute of Technology

Jones and Bartlett Publishers
Boston   *London*

Editorial, Sales, and Customer Service Offices
Jones and Bartlett Publishers
One Exeter Plaza
Boston, MA 02116

Jones and Bartlett Publishers International
P.O. Box 1498
London W6 7RS
England

Manuscript typed by Dan Robb

Library of Congress Cataloging-in-Publication Data

Morgan, Frank.
   Riemannian geometry : a beginner's guide / Frank Morgan.
      p.   cm.
   Includes bibliographical references and index.
   ISBN 0-86720-242-4
   1. Geometry, Riemannian. I. Title.
QA611.M674 1992
516.3'73—dc20   92-13261
                                               CIP

Printed in the United States of America

96 95 94 93 92 10 9 8 7 6 5 4 3 2 1

Photograph courtesy of the Morgan
family; taken by the author's
grandfather, Dr. Charles Selemeyer.

*This book is dedicated to my teachers—notably Fred
Almgren, Clem Collins, Arthur Mattuck, Mabel Riker,
my mom, and my dad. Here as a child I got an early
geometry lesson from my dad.*

F.M.

Contents

Preface

The complicated formulations of Riemannian geometry present a daunting aspect to the student. This little book focuses on the central concept—curvature. It gives a naive treatment of Riemannian geometry, based on surfaces in $\mathbf{R}^n$ rather than on abstract Riemannian manifolds.

The more sophisticated intrinsic formulas follow naturally. Later chapters treat hyperbolic geometry, general relativity, global geometry, and some current research on energy-minimizing curves and the isoperimetric problem. Proofs, when given at all, emphasize the main ideas and suppress the details that otherwise might overwhelm the student.

This book grew out of graduate courses I taught on tensor analysis at MIT in 1977 and on differential geometry at Stanford in 1987 and Princeton in 1990, and out of my own need to understand curvature better for my work. The last chapter includes research by Williams undergraduates. I want to thank my students, notably Alice Underwood; Paul Siegel, my teaching assistant for tensor analysis; and participants in a seminar at Washington and Lee led by Tim Murdoch.

Other books I have found helpful include Laugwitz's *Differential and Riemannian Geometry* [L], Hicks's *Notes on Differential Geometry* [Hi] (unfortunately out of print), Spivak's *Comprehensive Introduction to Differential Geometry* [S], and Stoker's *Differential Geometry* [St].

I am currently using this book and *Geometric Measure Theory: A Beginner's Guide* [M], both so happily edited by Klaus Peters and illustrated by Jim Bredt, as texts for an advanced, one-semester undergraduate course at Williams.

Williamstown   F.M.
Frank.Morgan@williams.edu

1
Introduction

The central concept of Riemannian geometry is *curvature*. It describes the most important geometric features of racetracks and of universes. We will begin by defining the curvature of a racetrack. Chapter 7 uses general relativity's interpretation of mass as curvature to predict the mysterious precession of Mercury's orbit.

The curvature κ of a racetrack is defined as the rate at which the direction vector $\mathbf{T}$ of motion is turning, measured in radians or degrees per meter. The curvature is big on sharp curves, zero on straightaways. See Figure 1.1.

A two-dimensional surface, such as the surface of Figure 1.2, can curve different amounts in different directions, perhaps upward in some directions, downward in others, and along straight lines in between. The principal curvatures κ_1 and κ_2 are the most upward (positive) and the most downward (negative), respectively. For the saddle of Figure 1.2, it appears that at the origin $\kappa_1 = \frac{1}{4}$ and $\kappa_2 = -1$. The *mean curvature* $H = \kappa_1 + \kappa_2 = -\frac{3}{4}$. The *Gauss curvature* $G = \kappa_1 \kappa_2 = -\frac{1}{4}$.

At the south pole of the unit sphere of Figure 1.3, $\kappa_1 = \kappa_2 = 1$, $H = 2$, and $G = 1$.

Since κ_1 and κ_2 measure the amount that the surface is curving in space, they could not be measured by a bug confined to the surface. They are "extrinsic" properties. Gauss made the astonishing discovery, however, that the Gauss curvature $G = \kappa_1 \kappa_2$ can, in prin-

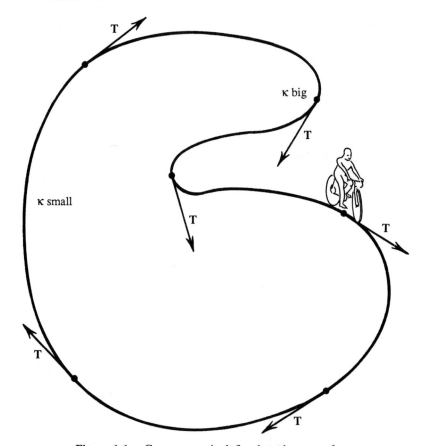

Figure 1.1. Curvature κ is defined as the rate of change of the direction vector **T**.

ciple, be measured from within the surface. This result, known as his *Theorema Egregium* or Remarkable Theorem, says that Gauss curvature is an "intrinsic" property.

An m-dimensional hypersurface in $\mathbf{R}^{m+1}$ has m principal curvatures $\kappa_1, \ldots, \kappa_m$ at each point. For an m-dimensional surface in $\mathbf{R}^n$, the situation is still more complicated; it is described not by numbers or by vectors, but by the second fundamental *tensor*. Still, Gauss's Theorema Egregium generalizes to show that an associated "Riemannian curvature tensor" is intrinsic.

Modern graduate texts in differential geometry strive to give intrinsic curvatures intrinsic definitions, which ignore the ambient $\mathbf{R}^n$

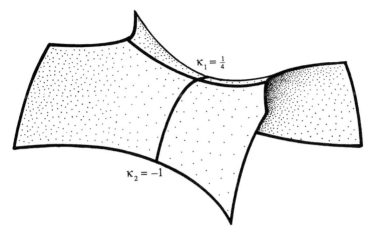

Figure 1.2. At the center of this saddle, the maximum upward curvature is $\kappa_1 = \frac{1}{4}$ and the maximum downward curvature is $\kappa_2 = -1$.

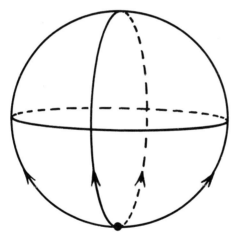

Figure 1.3. At the south pole, the curvature is $+1$ in all directions.

from the outset. In this text, surfaces will start out sitting in $\mathbf{R}^n$, where we can give concrete definitions of the second fundamental tensor and the Riemannian curvature tensor. Only later will we prove that the Riemannian curvature tensor actually is intrinsic.

2

Curves in $\mathbf{R}^n$

The central idea of Riemannian geometry — curvature — appears already for space curves in this chapter. For a parameterized curve $\mathbf{x}(t)$ in $\mathbf{R}^n$, with velocity $\mathbf{v} = \dot{\mathbf{x}}$ and unit tangent $\mathbf{T} = \mathbf{v}/|\mathbf{v}|$, the curvature vector $\boldsymbol{\kappa}$ is defined as the rate of change of $\mathbf{T}$ with respect to arc length:

$$\boldsymbol{\kappa} = d\,\mathbf{T}/ds = \frac{d\,\mathbf{T}/dt}{ds/dt} = \frac{1}{|\mathbf{v}|}\,\dot{\mathbf{T}}. \tag{1}$$

The curvature vector $\boldsymbol{\kappa}$ points in the direction in which $\mathbf{T}$ is turning, orthogonal to $\mathbf{T}$. Its length, the scalar curvature $\kappa = |\boldsymbol{\kappa}|$, gives the rate of turning. See Figure 2.1. For a planar curve with unit normal $\mathbf{n}$,

$$\kappa = |d\mathbf{n}/ds|. \tag{2}$$

For a circle of radius a, $\boldsymbol{\kappa}$ points toward the center, and $\kappa = 1/a$. For a general curve, the best approximating, or osculating, circle has radius $1/\kappa$, called the radius of curvature.

If the curve is parameterized by arc length, then the curvature vector $\boldsymbol{\kappa}$ simply equals $d^2\mathbf{x}/ds^2$. If the curve is the graph $\mathbf{y} = f(x)$ of a function $f: \mathbf{R} \to \mathbf{R}^{n-1}$ tangent to the x-axis at the origin 0, then

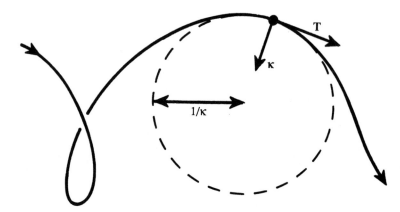

Figure 2.1. The curvature vector **κ** tells us which way the unit tangent vector **T** is turning and how fast. Its length |**κ**| is the reciprocal of the radius of the osculating circle.

$$\boldsymbol{\kappa}(0) = f''(0) \in \mathbf{R}^{n-1} \subset \mathbf{R} \times \mathbf{R}^{n-1}.$$

Without the tangency hypothesis, the scalar curvature

$$\kappa(0) = |f''(0)|/(1 + |f'(0)|^2)^{3/2}.$$

Curvature tells how the length of a curve changes as the curve is deformed. If an infinitesimal piece of planar curve ds is pushed a distance du in the direction of **κ**, the length changes by a factor of $1 - \kappa\, du$. Indeed, the original arc lies to second order on a circle of radius $1/\kappa$, and the new one on a circle of radius $1/\kappa - du = (1/\kappa)(1 - \kappa\, du)$. See Figure 2.2. More generally, if the displacement is a vector $d\mathbf{u}$ not necessarily in the direction of **κ**, only the component of $d\mathbf{u}$ in the **κ** direction matters, and the length changes by a factor of $1 - \boldsymbol{\kappa} \cdot d\mathbf{u}$. Hence the initial rate of change of length of a curve C in $\mathbf{R}^n$ with initial velocity $\mathbf{V} = d\mathbf{u}/dt$ is $-\int \boldsymbol{\kappa} \cdot \mathbf{V}\, ds$ (see Section 10.4).

2.1. The smokestack problem.   One day I got a call from a company constructing a huge smokestack, which required the attachment of a spiraling strip, or *strake*, for structural support (see Figures 2.3 and 2.4). Of course they had to cut the strake pieces out of a flat piece of metal (see Figure 2.5). The question was, What choice of inner radius r would make the strake fit on the smokestack best?

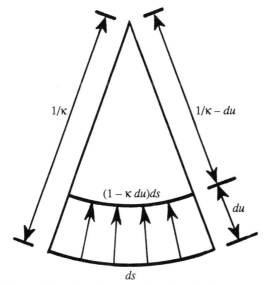

Figure 2.2. An element of arc length *ds* pushed in the direction of **κ** decreases by a factor of $1 - \kappa\, du$.

Figure 2.3. The metal strip, or strake, spirals around the smokestack on a helical path.

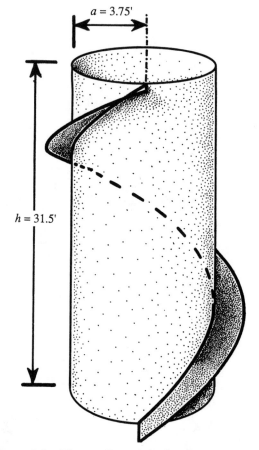

Figure 2.4. The smokestack had radius $a =$ 3.75 feet. Each revolution of the strake had a height of 31.5 feet.

The curve along which the strake attaches to the smokestack is a helix:

$$\mathbf{x} = (x, y, z) = (a \cos t, a \sin t, ht/2\pi),$$

with the x and y coordinates following a circle of radius $a = 3.75$ feet while the z coordinate increases at a constant rate. In each revolution, θ increases by 2π and z increases by $h = 31.5$ feet. Hence the velocity is

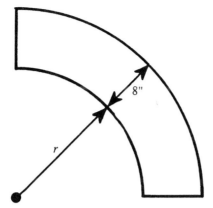

Figure 2.5. When pieces of strake are being cut out of a flat piece of metal, what inner radius r will make the strake fit best along the smokestack?

$$\mathbf{v} = \dot{\mathbf{x}} = (-a \sin t, a \cos t, h/2\pi),$$

and the speed is

$$\frac{ds}{dt} = |\mathbf{v}| = \sqrt{a^2 + \frac{h^2}{4\pi^2}} = c \approx 6.26 \text{ feet.}$$

The length of one revolution is

$$L = \int_0^{2\pi} |\mathbf{v}| \, dt = 2\pi c.$$

By analogy with a circle, an engineer guessed that the ideal inner cutting radius r would be $L/2\pi = c \approx 6.26$ feet. When he built a little model, however, he discovered that his guess was too small. After some trial and error, he found that strake pieces cut with $r \approx 10\frac{1}{2}$ feet fit well.

The way to compute the ideal r is to require the strake to have the right *curvature*. We will now compute the curvature κ of the helix and take r to be the radius of curvature $1/\kappa$ (that is, the radius of the circle with the same curvature).

The unit tangent vector $\mathbf{T} = \mathbf{v}/|\mathbf{v}| = \mathbf{v}/c$. Hence the curvature vector is

$$\kappa = d\mathbf{T}/ds = \frac{d\mathbf{T}/dt}{ds/dt} = \frac{\dot{\mathbf{v}}/c}{c} = \frac{1}{c^2}(-a\cos t, -a\sin t, 0),$$

and the scalar curvature is $\kappa = a/c^2$. Therefore the ideal inner radius is

$$r = 1/\kappa = c^2/a \approx 10.45 \text{ feet},$$

in close agreement with the engineer's experiment.

3

Surfaces in $\mathbf{R}^3$

This chapter studies the curvature of a C^2 surface $S \subset \mathbf{R}^3$ at a point $p \in S$. (C^2 just means that locally, the surface is the graph of a function with continuous second derivatives. Computing the curvature will involve differentiating twice.) Let T_pS denote the *tangent space* of vectors tangent to S at p. Let $\mathbf{n}$ denote a unit normal to S at p. *To study the curvature of S, we slice S by planes containing $\mathbf{n}$ and consider the curvature vector $\boldsymbol{\kappa}$ of the resulting curves.* (See Figure 3.1.) Of course each such $\boldsymbol{\kappa}$ must be a multiple of $\mathbf{n}$: $\boldsymbol{\kappa} = \kappa\mathbf{n}$. (For now we will allow κ to be positive or negative. The sign of κ depends on the choice of unit normal $\mathbf{n}$.) It will turn out that the largest and the smallest curvatures κ_1, κ_2 (called the principal curvatures) occur in orthogonal directions and determine the curvatures in all other directions.

Choose orthonormal coordinates on $\mathbf{R}^3$ with the origin at p, S tangent to the x, y-plane at p, and $\mathbf{n}$ pointing along the positive z-axis. Locally S is the graph of a function $z = f(x, y)$. Any unit vector $\mathbf{v}$ tangent to S at p, together with the unit normal vector $\mathbf{n}$, spans a plane, which intersects S in a curve. The curvature κ of this curve, which we call the curvature in the direction $\mathbf{v}$, is just the second derivative

$$\kappa = (D^2 f)_p(\mathbf{v}, \mathbf{v}) \equiv \mathbf{v}' \begin{bmatrix} \dfrac{\partial^2 f}{\partial x^2}(p) & \dfrac{\partial^2 f}{\partial x\, \partial y}(p) \\[2mm] \dfrac{\partial^2 f}{\partial x\, \partial y}(p) & \dfrac{\partial^2 f}{\partial y^2}(p) \end{bmatrix} \mathbf{v}.$$

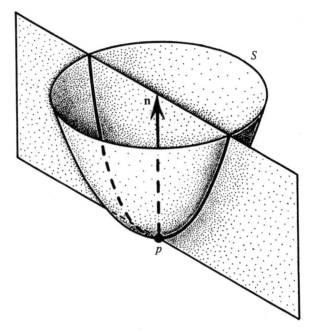

Figure 3.1. The curvature of a surface S at a point p is measured by the curvature of its slices by planes.

For example, if

$$\mathbf{v} = \begin{bmatrix} 1 \\ 0 \end{bmatrix}, \qquad \kappa = \frac{\partial^2 f}{\partial x^2}.$$

The bilinear form $(D^2 f)_p$ on $T_p S$ is called the *second fundamental form* II of S at p, given in coordinates as a symmetric 2×2 matrix:

$$\text{II} = D^2 f = \begin{bmatrix} \dfrac{\partial^2 f}{\partial x^2} & \dfrac{\partial^2 f}{\partial x\, \partial y} \\[2ex] \dfrac{\partial^2 f}{\partial x\, \partial y} & \dfrac{\partial^2 f}{\partial y^2} \end{bmatrix}.$$

This formula is good only at the point where the surface is tangent to the x,y-plane. For the second fundamental form, we will always use orthonormal coordinates.

Since II is symmetric, we may choose coordinates x, y such that II is diagonal:

$$II = \begin{bmatrix} \kappa_1 & 0 \\ 0 & \kappa_2 \end{bmatrix}.$$

Then the curvature κ in the direction $\mathbf{v} = (\cos\theta, \sin\theta)$ is given by Euler's formula (1760):

$$\kappa = II(\mathbf{v}, \mathbf{v}) = \mathbf{v}^T II \mathbf{v} = \kappa_1 \cos^2\theta + \kappa_2 \sin^2\theta,$$

a weighted average of κ_1 and κ_2. In particular, the largest and smallest curvatures are κ_1 and κ_2, obtained in the orthogonal directions we have chosen for the x- and y-axes.

3.1. Definitions.   At a point p in a surface $S \subset \mathbf{R}^3$, the eigenvalues κ_1, κ_2 of the second fundamental form II are called the *principal curvatures*, and the corresponding eigenvectors (uniquely determined unless $\kappa_1 = \kappa_2$) are called the *principal directions* or directions of curvature. The trace of II, $\kappa_1 + \kappa_2$, is called the *mean curvature H*. The determinant of II, $\kappa_1\kappa_2$, is called the *Gauss curvature G*.

Note that the signs of II and H but not of G depend on the choice of unit normal $\mathbf{n}$. Some treatments define the mean curvature as

$$\frac{1}{2}\text{trace }II = \frac{\kappa_1 + \kappa_2}{2}.$$

Just as the curvature κ gave the rate of change of the length of an evolving curve in Chapter 2, the mean curvature H gives the rate of change of the area of an evolving surface. Just as the rate of change of a function of several variables is called the directional derivative and depends on the direction of change, the initial rate of change of the area of a surface depends on its initial velocity $\mathbf{V}$ and is called the first variation.

3.2. Theorem.   *Let S be a C^2 surface in $\mathbf{R}^3$. The first variation of the area of S with respect to a compactly supported C^2 vectorfield $\mathbf{V}$ of S is given by integrating $\mathbf{V}$ against the mean curvature:*

$$\delta^1(S) = -\int_S \mathbf{V} \cdot H\mathbf{n}.$$

Remark. $\delta^1(S)$ is defined as

$$\frac{d}{dt}\,\text{area}\,(S + t\mathbf{V})\big|_{t=0},$$

or, equivalently,

$$\frac{d}{dt}\,\text{area}\,(f_t(S))\big|_{t=0},$$

where f_t is any C^3 deformation of space with initial velocity $\mathbf{V}$ on S. [$\delta^1(S)$ depends only on $\mathbf{V}$ and is linear in $\mathbf{V}$.] If S has infinite area, restrict attention to spt $\mathbf{V}$.

Proof.   Since the formula is linear in $\mathbf{V}$, we may consider tangential and normal variations separately. For tangential variations, which correspond to sliding the surface along itself, $\delta^1(S) = 0$, confirming the formula. Let $V\mathbf{n}$ be a small normal variation, and consider an infinitesimal square area $dx\,dy$ at p, where we may assume the principal directions point along the axes. To first order, the new infinitesimal area is

$$(1 - V\kappa_1)\,dx(1 - V\kappa_2)\,dy \approx (1 - VH)dx\,dy = (1 - \mathbf{V}\cdot H\mathbf{n})dx\,dy$$

(compare to Figure 2.2). The formula follows.

Remark.   A physical surface such as a soap film would tend to move in the normal direction of positive mean curvature in order to decrease its area, unless balanced by an opposite pressure. The mean curvature of a soap bubble in equilibrium is proportional to the pressure difference across it.

3.3. Minimal surfaces.   It follows from Theorem 3.2 that an area-minimizing surface, which minimizes area in competition with surfaces with the same boundary, must have vanishing mean curvature. Any surface with 0 mean curvature is called a *minimal surface*.

Some famous minimal surfaces are pictured in Figures 3.2 through 3.4. At each point, since the mean curvature vanishes, the principal curvatures must be equal in magnitude and opposite in sign.

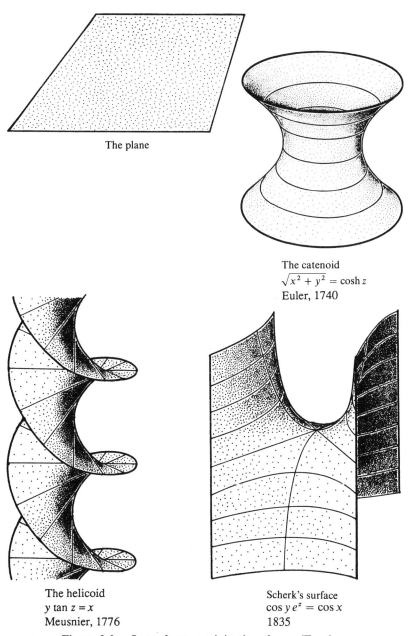

The plane

The catenoid
$\sqrt{x^2 + y^2} = \cosh z$
Euler, 1740

The helicoid
$y \tan z = x$
Meusnier, 1776

Scherk's surface
$\cos y \, e^z = \cos x$
1835

Figure 3.2. Some famous minimal surfaces. (Frank Morgan, *Geometric Measure Theory*, p. 68. © 1988, Academic Press. All rights reserved. Reprinted with permission of the publisher.)

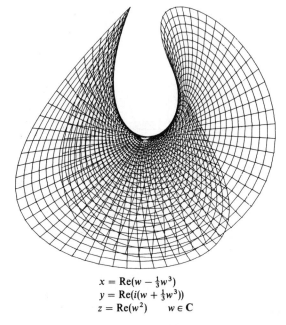

$$x = \mathrm{Re}(w - \tfrac{1}{3}w^3)$$
$$y = \mathrm{Re}(i(w + \tfrac{1}{3}w^3))$$
$$z = \mathrm{Re}(w^2) \qquad w \in \mathbf{C}$$

Figure 3.3. Enneper's surface 1864.

3.4. Coordinates, length, metric. Local coordinates or parameters u_1, u_2 on a C^2 surface $S \subset \mathbf{R}^3$ are provided by a C^2 diffeomorphism (or parameterization) between a domain in the u_1, u_2-plane and a portion of S.

For example, the standard spherical coordinates φ, θ provide local coordinates on all of the sphere of radius a except for the poles (where the longitude θ is undefined and φ is not differentiable). The position vector determined by these coordinates is

$$\mathbf{x} = (x, y, z) = (a \sin \varphi \cos \theta, a \sin \varphi \sin \theta, a \cos \varphi).$$

In general, the position is some function of the coordinates u_i. Along a curve, these coordinates are in turn functions of a single parameter t.

Subscripts on the position vector $\mathbf{x}$ will denote partial derivatives

Figure 3.4.   The newly discovered complete,
embedded minimal surface of Costa, Hoffman, and
Meeks [Cos], [HoM], [Ho]. (Courtesy of David
Hoffmann, Jim Hoffmann, and Michael Callahan.)

with respect to the u_i:

$$\mathbf{x}_i = \frac{\partial \mathbf{x}}{\partial u_i} = \left(\frac{\partial x}{\partial u_i}, \frac{\partial y}{\partial u_i}, \frac{\partial z}{\partial u_i} \right).$$

A dot will denote differentiation with respect to t:

$$\dot{\mathbf{x}} = \frac{d\mathbf{x}}{dt} = \sum \mathbf{x}_i \dot{u}_i \quad \text{(the chain rule)}.$$

The arc length of a curve in the surface with coordinates $u(t)$ is given by

$$L = \int_{t_0}^{t_1} |\dot{\mathbf{x}}| \, dt = \int_{t_0}^{t_1} |\mathbf{x}_1 \dot{u}_1 + \mathbf{x}_2 \dot{u}_2| \, dt$$

$$= \int_{t_0}^{t_1} \sqrt{(\mathbf{x}_1 \cdot \mathbf{x}_1) \dot{u}_1^2 + 2(\mathbf{x}_1 \cdot \mathbf{x}_2) \dot{u}_1 \dot{u}_2 + (\mathbf{x}_2 \cdot \mathbf{x}_2) \dot{u}_2^2} \, dt \qquad (1)$$

$$= \int_{t_0}^{t_1} \sqrt{\sum g_{ij} \dot{u}_i \dot{u}_j} \, dt,$$

where

$$g_{ij} \equiv \mathbf{x}_i \cdot \mathbf{x}_j \equiv \frac{\partial \mathbf{x}}{\partial u_i} \cdot \frac{\partial \mathbf{x}}{\partial u_j}. \qquad (2)$$

In other words, $L = \int ds$, where

$$ds^2 = \sum g_{ij} \, du_i \, du_j. \qquad (3)$$

For example, on the sphere of radius a, $L = \int ds$, where, as it turns out,

$$ds^2 = a^2 d\varphi^2 + a^2 \sin^2 \varphi \, d\theta^2 = (a^2 \dot{\varphi} + a^2 \sin^2 \varphi \dot{\theta}) \, dt^2,$$

so $g_{11} = a^2$, $g_{22} = a^2 \sin^2 \varphi$, and $g_{12} = g_{21} = 0$ (see Exercise 3.2).

The matrix $g = [g_{ij}]$ is called the *first fundamental form* or *metric*. It is an intrinsic quantity in that it relates to measurements inside the surface. Notice that in the formula for length,

$$\sum g_{ij} \dot{u}_i \dot{u}_j = g_{11} \dot{u}_1 \dot{u}_1 + g_{12} \dot{u}_1 \dot{u}_2 + g_{21} \dot{u}_2 \dot{u}_1 + g_{22} \dot{u}_2 \dot{u}_2$$
$$= g_{11} \dot{u}_1^2 + 2 g_{12} \dot{u}_1 \dot{u}_2 + g_{22} \dot{u}_2^2.$$

For many surfaces in $\mathbf{R}^3$, it is convenient to use x, y as local coordinates and consider $z(x, y)$. Then

$$\mathbf{x}_1 = (1, 0, z_x) \qquad \text{and} \qquad \mathbf{x}_2 = (0, 1, z_y).$$

The following proposition gives useful formulas for the mean curvature H and Gauss curvature G.

3.5. Proposition.   *For any local coordinates u_1, u_2 about a point p*

in a C^2 surface in $\mathbf{R}^3$, the second fundamental form II *at p is similar to*

$$g^{-1}(D^2\mathbf{x}) \cdot \mathbf{n} \equiv g^{-1}\begin{bmatrix} \mathbf{x}_{11} \cdot \mathbf{n} & \mathbf{x}_{12} \cdot \mathbf{n} \\ \mathbf{x}_{12} \cdot \mathbf{n} & \mathbf{x}_{22} \cdot \mathbf{n} \end{bmatrix},$$

where

$$\mathbf{x}_{ij} \equiv \frac{\partial^2 \mathbf{x}}{\partial u_i \, \partial u_j} \quad and \quad \mathbf{n} = \frac{\mathbf{x}_1 \times \mathbf{x}_2}{|\mathbf{x}_1 \times \mathbf{x}_2|}.$$

Consequently,

$$H = \text{trace } g^{-1}(D^2\mathbf{x}) \cdot \mathbf{n} \tag{1}$$

$$= \frac{\mathbf{x}_2^2 \mathbf{x}_{11} - 2(\mathbf{x}_1 \cdot \mathbf{x}_2)\mathbf{x}_{12} + \mathbf{x}_1^2 \mathbf{x}_{22}}{\mathbf{x}_1^2 \mathbf{x}_2^2 - (\mathbf{x}_1 \cdot \mathbf{x}_2)^2} \cdot \mathbf{n},$$

$$G = \det (g^{-1}(D^2\mathbf{x}) \cdot \mathbf{n}) \tag{2}$$

$$= \frac{(\mathbf{x}_{11} \cdot \mathbf{n})(\mathbf{x}_{22} \cdot \mathbf{n}) - (\mathbf{x}_{12} \cdot \mathbf{n})^2}{\mathbf{x}_1^2 \mathbf{x}_2^2 - (\mathbf{x}_1 \cdot \mathbf{x}_2)^2}.$$

Before turning to the proof of Proposition 3.5, we note that

(A) Given H and G, you can solve easily for the principal curvatures:

$$\kappa = \frac{H \pm \sqrt{H^2 - 4G}}{2}.$$

(B) If the surface is a graph $\mathbf{x} = (x, y, f(x, y))$, then

$$H = \frac{(1 + f_y^2)f_{xx} - 2f_x f_y f_{xy} + (1 + f_x^2)f_{yy}}{(1 + f_x^2 + f_y^2)^{3/2}}, \tag{3}$$

$$G = \frac{f_{xx}f_{yy} - f_{xy}^2}{(1 + f_x^2 + f_y^2)^2}. \tag{4}$$

Proof. We may assume that S is tangent to the x, y-plane at $p = 0$, so S is locally a graph $z = f(x, y)$ with $f_x(0) = f_y(0) = 0$ and $\mathbf{n}(0) = (0, 0, 1)$. For the particular local coordinates x, y,

$$\mathbf{x} = (x, y, f(x, y)), \qquad g(0) = I.$$

The proposition says that II is similar to

$$\begin{bmatrix} f_{xx} & f_{xy} \\ f_{xy} & f_{yy} \end{bmatrix}_0,$$

which is correct; indeed, they are equal.

Now let u_1, u_2 be any local coordinates, and let J denote the Jacobian at 0:

$$J = \begin{bmatrix} \dfrac{\partial x}{\partial u_1} & \dfrac{\partial x}{\partial u_2} \\ \dfrac{\partial y}{\partial u_1} & \dfrac{\partial y}{\partial u_2} \end{bmatrix}_0.$$

Since $\left.\dfrac{\partial z}{\partial u_1}\right|_0 = \left.\dfrac{\partial z}{\partial u_2}\right|_0 = 0,$

$$g = J^T J.$$

Then by the chain rule, since $\dfrac{\partial \mathbf{x}}{\partial x} \cdot \mathbf{n} = 0$ and $\dfrac{\partial \mathbf{x}}{\partial y} \cdot \mathbf{n} = 0,$

$$g^{-1}(D^2\mathbf{x}) \cdot \mathbf{n} = (J^T J)^{-1} J^T \begin{bmatrix} \dfrac{\partial^2 \mathbf{x}}{\partial x^2} \cdot \mathbf{n} & \dfrac{\partial^2 \mathbf{x}}{\partial x\, \partial y} \cdot \mathbf{n} \\ \dfrac{\partial^2 \mathbf{x}}{\partial x\, \partial y} \cdot \mathbf{n} & \dfrac{\partial^2 \mathbf{x}}{\partial y^2} \cdot \mathbf{n} \end{bmatrix} J = J^{-1} \mathrm{II} J$$

is indeed similar to II.

Example. We will compute the curvature of the catenoid $\sqrt{x^2 + y^2} = \cosh z$ of Figure 3.2. At most points we could use x and y as coordinates. Instead, we will use z and the polar coordinate θ. The equation says that $r = \cosh z$. Hence the position

$$\mathbf{x} = (x, y, z) = (\cosh z \cos \theta, \cosh z \sin \theta, z)$$

$$\mathbf{x}_1 = (\sinh z \cos \theta, \sinh z \sin \theta, 1)$$

$$\mathbf{x}_2 = (-\cosh z \sin \theta, \cosh z \cos \theta, 0)$$

$$\mathbf{x}_{11} = (\cosh z \cos \theta, \cosh z \sin \theta, 0)$$

$$\mathbf{x}_{12} = (-\sinh z \sin \theta, \sinh z \cos \theta, 0)$$

$$\mathbf{x}_{22} = (-\cosh z \cos \theta, -\cosh z \sin \theta, 0)$$

$$\mathbf{n} = \frac{(\cos\theta,\ \sin\theta,\ \sinh z)}{\cosh z}.$$

By Proposition 3.5,

$$H = \frac{\cosh^2 z\ \mathbf{x}_{11} - 0 + \cosh^2 z\ \mathbf{x}_{22}}{\text{something}} \cdot \mathbf{n} = 0,$$

so the catenoid is indeed a minimal surface and $\kappa_1 = -\kappa_2$.

$$G = \frac{(1)(-1) - 0}{\cosh^2 z\ \cosh^2 z - 0} = -\cosh^{-4} z.$$

Hence $\kappa_1 = -\kappa_2 = \cosh^{-2} z$.

3.6. Gauss's Theorema Egregium. *Gauss curvature G is intrinsic. Specifically, there are local coordinates u_1, u_2 about any point p in a C^3 surface S in $\mathbf{R}^3$ such that the first fundamental form g at p is I to first order. In any such coordinate system, the Gauss curvature is*

$$G = \frac{\partial^2 g_{12}}{\partial u_1\ \partial u_2} - \frac{1}{2}\frac{\partial^2 g_{22}}{\partial u_1^2} - \frac{1}{2}\frac{\partial^2 g_{11}}{\partial u_2^2}.$$

Remark. To say that $G = I$ to first order means that $g_{11}(p) = g_{22}(p) = 1$, $g_{12}(p) = 0$, and each

$$g_{ij,k}(p) = \frac{\partial g_{ij}}{\partial u_k}(p) = 0.$$

Proof. Locally S is the graph of a function f over its tangent plane. Orthonormal coordinates on the tangent plane make the metric g equal to I to first order. We may assume that S is tangent to the x, y-plane at $p = 0$. In x, y coordinates,

$$g = \begin{bmatrix} 1 + f_x^2 & f_x f_y \\ f_x f_y & 1 + f_y^2 \end{bmatrix},$$

and one computes that at 0

$$\frac{\partial^2 g_{12}}{\partial x\ \partial y} - \frac{1}{2}\frac{\partial^2 g_{22}}{\partial x^2} - \frac{1}{2}\frac{\partial^2 g_{11}}{\partial y^2} = f_{xx}f_{yy} - f_{xy}^2 = \det D^2 f = \det \mathrm{II} = G.$$

Any coordinates for which the metric at p is I to first order agree with orthonormal coordinates on the tangent plane to first order and hence yield the same result.

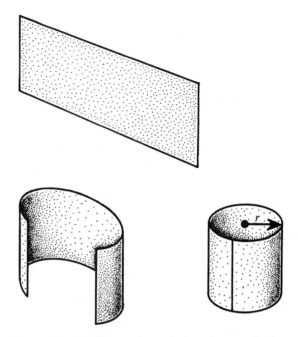

Figure 3.5. Rolling a piece of plane into a cylinder of radius r changes the principal curvatures κ_1, κ_2 from 0, 0 to $1/r$, 0 and changes the mean curvature $H = \kappa_1 + \kappa_2$ from 0 to $1/r$. The Gauss curvature, however, remains $G = \kappa_1\kappa_2 = 0$, as Gauss's Theorema Egregium guarantees.

Remark. A full appreciation of Gauss's Theorema Egregium requires the realization that most extrinsic quantities are not intrinsic.

In $\mathbf{R}^3$, a flat piece of plane can be rolled, or "bent," into a piece of cylinder of radius r without changing anything a bug on the surface could detect. This bending, however, does change the curvature κ of an arc of latitude of the cylinder from 0 in the plane to $1/r$ on the cylinder; does change the principal curvatures κ_1, κ_2 from 0, 0 to $1/r$, 0; and does change the mean curvature H from 0 to $1/r$. The Gauss curvature, however, remains $G = \kappa_1\kappa_2 = 0$, as the theorem guarantees. See Figure 3.5.

No kind of curvature can be detected by a bug on a curve. But if the bug moves to a surface, it can detect Gaussian curvature.

3.7. Gauss curvature and area.   An intrinsic definition of the Gauss curvature G at a point p in a surface could be based on the formula for the area of a disc of intrinsic radius r about p:

$$\text{area} = \pi r^2 - G \frac{\pi}{12} r^4 + \cdots . \tag{1}$$

Other interpretations of G will appear in Sections 8.1 and 8.6.

EXERCISES

3.1. What are the principal curvatures κ_1, κ_2, the Gauss curvature G, and the mean curvature H at each of the following?
   a. At a point on a sphere of radius a
   b. At the origin for the graph $z = f(x, y) = ax^2 + by^2$
   c. At the origin for the graph $z = f(x, y) = 66x^2 - 24xy + 59y^2$
   d. At the origin for the graph $z = f(x, y) = x + 2x^2 + 3y^2$
   e. At a general point on the helicoid $y \tan z = x$
   f. At a general point on the ellipsoid $9x^2 + 4y^2 + z^2 = 36$

3.2. For the standard coordinates $u_1 = \theta$, $u_2 = \varphi$ on the sphere of radius a, compute the first fundamental form $[g_{ij}]$. Use it to calculate the length of a circle of latitude $\varphi = c$.

3.3. Derive formulas 3.5(3) and 3.5(4) for the curvatures of a graph from 3.5(1) and 3.5(2).

3.4. Obtain a surface of revolution in $\mathbf{R}^3$ by revolving a curve $x = f(z)$ in the x,z-plane about the z-axis. Check that the surface is parametrized by cylindrical coordinates θ, z as

$$\mathbf{x} = (f(z) \cos \theta, \ f(z) \sin \theta, z).$$

Use Proposition 3.5 to show that the inward mean curvature is given by

$$H = \kappa + \frac{1}{f\sqrt{1 + f'^2}}, \tag{$*$}$$

where κ is the inward curvature of the original curve ($\kappa = -(f''/(1 + f'^2)^{3/2})$). Check formula ($*$) for a sphere centered at the origin.

3.5. Do the first-year calculus exercise in spherical coordinates of computing the area of a polar cap of intrinsic radius r on a sphere of radius a to obtain

$$\text{area} = 2\pi a^2 \left(1 - \cos \frac{r}{a} \right).$$

Then verify equation 3.7(1) for this case.

4

Surfaces in $\mathbf{R}^n$

This chapter shows how the theory of curvature at a point p in a 2-dimensional surface S extends from $\mathbf{R}^3$ to $\mathbf{R}^n$. As before, choose orthonormal coordinates on $\mathbf{R}^n$ with the origin at p and S tangent to the x_1, x_2-plane at p. Locally S is the graph of a function

$$f: T_p S \to T_p S^\perp.$$

Any unit vector v tangent to S at p, together with the vectors normal to S at p, spans a hyperplane, which intersects S in a curve. The curvature vector κ of this curve, which we call the curvature in the direction v, is just the second derivative

$$\kappa = (D^2 f)_p(v, v).$$

(We will soon have so many tangent vectors v around that we are now abandoning boldface notation $\mathbf{v}$ for them.)

The bilinear form $(D^2 f)_p$ on $T_p S$ with values in $T_p S^\perp$ is called the *second fundamental tensor* II of S at p, given in coordinates as a symmetric 2×2 matrix with entries in $T_p S^\perp$:

149,752

$$\mathrm{II} = \begin{bmatrix} \dfrac{\partial^2 f}{\partial x_1^2} & \dfrac{\partial^2 f}{\partial x_1\, \partial x_2} \\[2ex] \dfrac{\partial^2 f}{\partial x_1\, \partial x_2} & \dfrac{\partial^2 f}{\partial x_2^2} \end{bmatrix} = \begin{bmatrix} a_{11} & a_{12} \\ a_{12} & a_{22} \end{bmatrix}.$$

Again, this formula is good only at the point where the surface is tangent to the x_1,x_2-plane. If $n = 3$, this second fundamental tensor is just the second fundamental form times the unit normal $\mathbf{n}$.

Generally this matrix cannot be diagonalized to produce principal curvatures. The trace of II, $a_{11} + a_{22} \in T_pS^{\perp}$, is called the *mean curvature vector* $\mathbf{H}$. (If $n = 3$, $\mathbf{H} = H\mathbf{n}$.) The scalar quantity $a_{11} \cdot a_{22} - a_{12} \cdot a_{12}$ is called the *Gauss curvature G*. Neither $\mathbf{H}$ nor G depends on the choice of orthonormal coordinates.

Let G_i denote the Gauss curvature of the projection S_i of S into

$$\mathbf{R}_1 \times \mathbf{R}_2 \times \{0\} \times \cdots \times \mathbf{R}_i \times \cdots \times \{0\} \equiv \mathbf{R}_i^3.$$

Then the Gauss curvature G of S at p is

$$G = \sum_{i=3}^{n} G_i$$

simply because the dot product of two vectors is just the sum of the products of the components.

4.1. Theorem.   *Let S be a C^2 surface in $\mathbf{R}^n$. The first variation of the area of S with respect to a compactly supported C^2 vectorfield $\mathbf{V}$ on S is given by integrating $\mathbf{V}$ against the mean curvature vector:*

$$\delta^1(S) = -\int_S \mathbf{V} \cdot \mathbf{H}.$$

Proof.   Since the formula is linear in $\mathbf{V}$, we may consider variations in the $x_1, x_2, \ldots$ directions separately. For the x_1, x_2 directions, which correspond to sliding the surface along itself, $\delta^1(S) = 0$, as the formula says. Let $\mathbf{V}$ be a small variation in the x_3 direction, and consider an infinitesimal square area $dx_1\, dx_2$ at p, where we may assume that the x_3 component of II is diagonal:

$$\begin{bmatrix} \kappa_1 & 0 \\ 0 & \kappa_2 \end{bmatrix}.$$

To first order, it is displaced to an infinitesimal area

$$(1 \mp |\mathbf{V}|\kappa_1)\, dx_1\, (1 \mp |\mathbf{V}|\kappa_2)\, dx_2 \approx (1 - \mathbf{V} \cdot \mathbf{H})dx_1\, dx_2.$$

The formula follows.

The concepts of local coordinates and the first fundamental form extend without change from $\mathbf{R}^2$ to $\mathbf{R}^n$. Likewise Proposition 3.5 generalizes as follows.

4.2. Proposition.   *For any local coordinates* u_1, u_2 *about a point* p *in a* C^2 *surface* S *in* $\mathbf{R}^n$, *the second fundamental tensor* II *at* p *is similar to*

$$g^{-1}P(D^2\mathbf{x}) \equiv g^{-1}\begin{bmatrix} P(\mathbf{x}_{11}) & P(\mathbf{x}_{12}) \\ P(\mathbf{x}_{12}) & P(\mathbf{x}_{22}) \end{bmatrix},$$

where P *denotes projection onto* $T_pS^{\perp}$ *and*

$$\mathbf{x}_{ij} \equiv \frac{\partial^2 \mathbf{x}}{\partial u_i\, \partial u_j}.$$

Consequently,

$$\mathbf{H} = \text{trace } g^{-1}P(D^2\mathbf{x}) \tag{1}$$

$$= P\frac{\mathbf{x}_2^2\mathbf{x}_{11} - 2(\mathbf{x}_1 \cdot \mathbf{x}_2)\mathbf{x}_{12} + \mathbf{x}_1^2\mathbf{x}_{22}}{\mathbf{x}_1^2\mathbf{x}_2^2 - (\mathbf{x}_1 \cdot \mathbf{x}_2)^2}$$

$$G = \det\left(g^{-1}P(D^2\mathbf{x})\right) \tag{2}$$

$$= \frac{(P\mathbf{x}_{11}) \cdot (P\mathbf{x}_{22}) - (P\mathbf{x}_{12})^2}{\mathbf{x}_1^2\mathbf{x}_2^2 - (\mathbf{x}_1 \cdot \mathbf{x}_2)^2}.$$

Remark. $T_pS^{\perp}$ *and hence* P *change from point to point. If* $T_pS^{\perp}$ *is just the* $x_3, \ldots, x_n$-*plane, then*

$$P(a_1, a_2, a_3, a_4, \ldots) = (0, 0, a_3, a_4, \ldots).$$

If $\mathbf{x}_1$, $\mathbf{x}_2$ give an *orthogonal* basis for T_pS, then

$$P(\mathbf{w}) = \mathbf{w} - \frac{\mathbf{w} \cdot \mathbf{x}_1}{\mathbf{x}_1 \cdot \mathbf{x}_1}\mathbf{x}_1 - \frac{\mathbf{w} \cdot \mathbf{x}_2}{\mathbf{x}_2 \cdot \mathbf{x}_2}\mathbf{x}_2. \tag{3}$$

If $\mathbf{x}_1$, $\mathbf{x}_2$ are not orthogonal, compute P by replacing $\mathbf{x}_2$ by

$$\mathbf{x}_2 - \frac{\mathbf{x}_2 \cdot \mathbf{x}_1}{\mathbf{x}_1 \cdot \mathbf{x}_1}\mathbf{x}_1.$$

Example.   Consider the surface

$$\{(w, z) \in \mathbf{C}^2 : w = e^z\}.$$

We will use $x = Re\ z$ and $y = Im\ z$ as coordinates. Then

$$\mathbf{x} = (x, y, e^x \cos y, e^x \sin y)$$
$$\mathbf{x}_1 = (1, 0, e^x \cos y, e^x \sin y)$$
$$\mathbf{x}_2 = (0, 1, -e^x \sin y, e^x \cos y)$$
$$\mathbf{x}_{11} = (0, 0, e^x \cos y, e^x \sin y)$$
$$\mathbf{x}_{12} = (0, 0, -e^x \sin y, e^x \cos y)$$
$$\mathbf{x}_{22} = (0, 0, -e^x \cos y, -e^x \sin y)$$

Note that $\mathbf{x}_1^2 = \mathbf{x}_2^2 = 1 + e^{2x}$, $\mathbf{x}_1 \cdot \mathbf{x}_2 = 0$. Hence,

$$H = P\frac{(0, 0, 0, 0)}{(1 + e^{2x})^2} = \mathbf{0}.$$

This is a minimal surface. (As a matter of fact, every complex analytic variety is a minimal surface.)

To compute G, first compute $P(\mathbf{x}_{ij})$. Since $\mathbf{x}_1$, $\mathbf{x}_2$ are orthogonal,

$$P(\mathbf{x}_{11}) = \mathbf{x}_{11} - \frac{\mathbf{x}_{11} \cdot \mathbf{x}_1}{\mathbf{x}_1 \cdot \mathbf{x}_1}\mathbf{x}_1 - \frac{\mathbf{x}_{11} \cdot \mathbf{x}_2}{\mathbf{x}_2 \cdot \mathbf{x}_2}\mathbf{x}_2$$
$$= \frac{(-e^{2x}, 0, e^x \cos y, e^x \sin y)}{1 + e^{2x}}.$$

Similarly,

$$P(\mathbf{x}_{12}) = \frac{(0, -e^{2x}, -e^x \sin y, e^x \cos y)}{1 + e^{2x}},$$

$$P(\mathbf{x}_{22}) = \frac{(e^{2x}, 0, -e^x \cos y, -e^x \sin y)}{1 + e^{2x}}.$$

Hence,

$$G = \frac{(-e^{4x} - e^{2x}) - (e^{4x} + e^{2x})}{(1 + e^{2x})^2[(1 + e^{2x})^2 - 0]} = -\frac{2e^{2x}}{(1 + e^{2x})^3}.$$

4.3. Gauss's Theorema Egregium. Finally, Gauss's Theorema Egregium, with the same proof as in Section 3.6, says that G is intrinsic, given in local coordinates u_1, u_2 in which $G = I$ to first order by the formula

$$G = \frac{\partial^2 g_{12}}{\partial u_1 \partial u_2} - \frac{1}{2}\frac{\partial^2 g_{22}}{\partial u_1^2} - \frac{1}{2}\frac{\partial^2 g_{11}}{\partial u_2^2}.$$

EXERCISES

4.1. Compute the mean curvature vector and the Gauss curvature at each of the following:

a. At the origin for the graph

$$(z, w) = f(x, y) = (x^2 + 2y^2, 66x^2 - 24xy + 59y^2).$$

[Then compare with Exercise 3.1(b, c).]

b. At a general point of $\{(z, w) \in \mathbf{C}^2 : w = z^2\}$.

4.2. Show that for the graph of a complex analytic function f,

$$\{w = f(z)\} \subset \mathbf{C}^2,$$

$$\mathbf{H} = \mathbf{0},$$

and

$$G = -2|f''(z)|^2(1 + |f'(z)|^2)^{-3}.$$

In particular, the graph of a complex analytic function is a minimal surface. (Compare with the example after Proposition 4.2.)

4.3. *Minimal surface equation.* Show that the graph of a function $f : \mathbf{R}^2 \to \mathbf{R}^{n-2}$ is a minimal surface if and only if

$$(1 + |f_y|^2)f_{xx} - 2(f_x \cdot f_y)f_{xy} + (1 + |f_x|^2)f_{yy} = \mathbf{0}.$$

[Compare with formula 3.5(3) for the special case $n = 3$.]

5

m-Dimensional Surfaces in $\mathbf{R}^n$

This chapter extends the theory to C^2 m-dimensional surfaces S in $\mathbf{R}^n$. As before, choose orthonormal coordinates on $\mathbf{R}^n$ with the origin at p and S tangent to the $x_1, x_2, \ldots, x_m$-plane at p. Locally S is the graph of a function

$$f: T_p S \to T_p S^{\perp}.$$

A unit vector v tangent to S at p, together with the vectors normal to S at p, spans a plane, which intersects S in a curve. The curvature vector $\boldsymbol{\kappa}$ of the curve, which we call the curvature in the direction v, is just the second derivative

$$\boldsymbol{\kappa} = (D^2 f)_p(v, v).$$

The bilinear form $(D^2 f)_p$ on $T_p S$ with values in $T_p S^{\perp}$ is called the *second fundamental tensor* II of S at p, given in coordinates as a symmetric $m \times m$ matrix with entries in $T_p S^{\perp}$:

$$\mathrm{II} = \begin{bmatrix} \dfrac{\partial^2 f}{\partial x_1^2} & & \dfrac{\partial^2 f}{\partial x_1 \, \partial x_m} \\ & \ddots & \\ \dfrac{\partial^2 f}{\partial x_1 \, \partial x_m} & & \dfrac{\partial^2 f}{\partial x_m^2} \end{bmatrix}.$$

The trace of II is called the *mean curvature vector* **H**. [Some treatments define **H** as (trace II)/n.]

   Hypersurfaces.   For hypersurfaces ($n = m + 1$), II is just the unit normal **n** times a scalar matrix, called the second fundamental form and also denoted by II. **H** = H**n**, where H is the (*scalar*) *mean curvature*. If we choose coordinates to make the second fundamental form diagonal,

$$II = \begin{bmatrix} \kappa_1 & & & 0 \\ & \cdot & & \\ & & \cdot & \\ 0 & & & \kappa_m \end{bmatrix},$$

then $H = \kappa_1 + \cdots + \kappa_m$. If the unit normal **n** $= (n_1, \ldots, n_n)$ is extended locally as a unit vectorfield, then $\partial n_n / \partial x_n = 0$, while for $1 \le i \le n - 1$, $\partial n_i / \partial x_i = -\kappa_i$ [compare to equation 2.0(2)]. Hence

$$H = - \sum_{i=1}^{n} \partial n_i / \partial x_i \equiv -\text{div } \mathbf{n}.$$

If the hypersurface is given as a level set $\{f(x_1, \ldots, x_n) = c\}$, then **n** $= \nabla f / |\nabla f|$, where $\nabla f \equiv (\partial n_1 / \partial x_1, \ldots, \partial n_n / \partial x_n)$, and

$$H = -\text{div} \frac{\nabla f}{|\nabla f|}. \tag{1}$$

5.1. Theorem.   *Let S be a C^2 m-dimensional surface in $\mathbf{R}^n$. The first variation of the area of S with respect to a compactly supported C^2 vectorfield $\mathbf{V}$ on S is given by integrating $\mathbf{V}$ against the mean curvature vector:*

$$\delta^1(S) = - \int_S \mathbf{V} \cdot \mathbf{H}.$$

   Proof.   Since the formula is linear in **v**, we may consider variations in the $x_1, x_2, \ldots, x_n$ directions separately. For the $x_1, \ldots, x_m$ directions, which correspond to sliding the surface along itself, $\delta^1(S) = 0$, as the formula says. Let **V** be a small variation in the x_j direction ($m < j \le n$), and consider an infinitesimal area $dx_1 \cdots dx_m$ at p, where we may assume that the x_j component of II is diagonal:

$$\begin{bmatrix} \kappa_1 & & 0 \\ & \cdot & \\ & & \cdot \\ 0 & & \kappa_m \end{bmatrix}.$$

To first order, it is displaced to an infinitesimal area

$$(1 \mp |\mathbf{V}|\kappa_1)\, dx_1 \cdots (1 \mp |\mathbf{V}|\kappa_m)\, dx_m \approx (1 - \mathbf{V} \cdot \mathbf{H})dx_1 \cdots dx_m.$$

The formula follows.

5.2. Sectional and Riemannian curvature. The sectional curvature K of S at p assigns to every 2-plane $P \subset T_pS$ the Gauss curvature of the 2-dimensional surface

$$S \cap (P \oplus T_pS^\perp).$$

If v, w give an orthonormal basis for P, then by its definition the sectional curvature is

$$K(P) = \mathrm{II}(v, v) \cdot \mathrm{II}(w, w) - \mathrm{II}(v, w) \cdot \mathrm{II}(v, w). \tag{1}$$

For example, if $\mathrm{II} = [a_{ij}]$ and $P = e_1 \wedge e_2$ is the x_1,x_2-plane, then the sectional curvature is

$$\begin{aligned} K(P) &= \mathrm{II}(e_1, e_1) \cdot \mathrm{II}(e_2, e_2) - \mathrm{II}(e_1, e_2) \cdot \mathrm{II}(e_1, e_2) \\ &= a_{11} \cdot a_{22} - a_{12} \cdot a_{12}. \end{aligned}$$

Remark. For hypersurfaces ($n = m + 1$), for any 2-plane $P = \Sigma\, p_{ij}\, e_i \wedge e_j$, if we choose coordinates to make the second fundamental form diagonal,

$$\mathrm{II} = \begin{bmatrix} \kappa_1 & & 0 \\ & \cdot & \\ & & \cdot \\ 0 & & \kappa_m \end{bmatrix},$$

then

$$K(P) = \sum_{1 \le i < j \le m} p_{ij}^2 \kappa_i \kappa_j.$$

Thus any sectional curvature $K(P)$ is a weighted average of the sectional curvatures $\kappa_i \kappa_j$ of the axis 2-planes $e_i \wedge e_j$.

For $2 < m < n$, $\mathbf{R}^n \cong T_pS \times \mathbf{R}_1 \times \cdots \times \mathbf{R}_{n-m}$, let S_i denote the projection of S into $T_pS \times \mathbf{R}_i$, with sectional curvature K_i. Then, by (1), the sectional curvature K of S satisfies $K = \Sigma\, K_i$.

Hence the sectional curvature of an m-dimensional surface S in $\mathbf{R}^n$ may be computed by separately diagonalizing the $n - m$ components of II, taking the appropriate weighted average of products of principal curvatures for each component, and summing over all components.

If $m = n - 1$, then II is a symmetric bilinear form called the second fundamental form. Its eigenvalues $\kappa_1, \ldots, \kappa_m$ are called the principal curvatures. Since $(D^2 f)_p$ is symmetric, in some orthogonal coordinates it is diagonal and f takes the form

$$f = \frac{\kappa_1 x_1^2}{2} + \frac{\kappa_2 x_2^2}{2} + \cdots + \frac{\kappa_m x_m^2}{2} + o(\mathbf{x}^2).$$

In general, if $\text{II} = (a_{ij})$, then formula (1) yields

$$K(P) = \left(\sum a_{ik} v_i v_k \right) \cdot \left(\sum a_{jl} w_j w_l \right) - \left(\sum a_{jk} v_k w_j \right) \cdot \left(\sum a_{il} v_i w_l \right)$$

$$= \sum R_{ijkl} v_i w_j v_k w_l, \tag{2}$$

where

$$R_{ijkl} = a_{ik} \cdot a_{jl} - a_{jk} \cdot a_{il} \tag{3}$$

are the 2×2 minors of II, corresponding to rows i, j and columns k, l. For example, $R_{1234} = a_{13} \cdot a_{24} - a_{14} \cdot a_{23}$ comes from rows 1, 2 and columns 3, 4 of

$$\text{II} = \begin{bmatrix} a_{11} & a_{12} & a_{13} & a_{14} & \cdots \\ a_{12} & a_{22} & a_{23} & a_{24} & \cdots \\ \vdots & \vdots & \vdots & \vdots & \end{bmatrix}.$$

$R_{1212} = a_{11} \cdot a_{22} - a_{12} \cdot a_{12}$ is the sectional curvature of the x_1, x_2-plane.

R is called the *Riemannian curvature tensor*. Thus, the Riemannian curvature tensor is just the 2×2 minors of the second fundamental tensor.

Immediately,

$$R_{jikl} = R_{ijlk} = -R_{ijkl} \tag{4}$$

(interchanging two rows or columns changes the sign of the minor), and

$$R_{klij} = R_{ijkl} \tag{5}$$

because II is symmetric. One can easily check Bianchi's first identity on permutation of the last three indices:

$$R_{ijkl} + R_{iklj} + R_{iljk} = 0. \tag{6}$$

To obtain a definition of R independent of the choice of orthonormal coordinates on T_pS, note that R is the bilinear form II $\wedge$ II on $\bigwedge^2 T_pS$. Indeed, if $\{e_i\}$ gives a basis for T_pS, so that $\{e_k \wedge e_l : k < l\}$ gives a basis for $\bigwedge^2 T_pS$, then

$$\text{II} \wedge \text{II}(e_k \wedge e_l) \equiv \text{II}(e_k) \wedge \text{II}(e_l) = \left(\sum a_{rk}e_r\right) \wedge \left(\sum a_{sl}e_s\right),$$

and

$$(e_i \wedge e_j) \cdot \text{II} \wedge \text{II}(e_k \wedge e_l) = a_{ik} \cdot a_{jl} - a_{jk} \cdot a_{il} = R_{ijkl}.$$

As a bilinear form on $\bigwedge^2 T_pS$, R is characterized by the values $\zeta \cdot R(\zeta)$ for unit 2-vectors $\zeta \in \bigwedge^2 T_pS$. Actually R is determined by the sectional curvatures $P \cdot R(P)$ for 2-planes (unit *simple* 2-vectors).

The *Ricci curvature* Ric is a bilinear form on T_pS, defined as a kind of trace of the Riemannian curvature. Just as the trace of a matrix $[c_{ij}]$ is a sum Σc_{ii} over a repeated subscript, the coordinates R_{jk} of the Ricci curvature are given by

$$R_{jl} = \sum_i R_{ijil}. \tag{7}$$

If you think of R_{ijkl} as a matrix of matrices,

$$\begin{bmatrix} [R_{i1k1}] & [R_{i1k2}] & \cdots & [R_{i1km}] \\ \vdots & \vdots & \vdots & \vdots \\ [R_{imk1}] & [R_{imk2}] & \cdots & [R_{imkm}] \end{bmatrix},$$

then R_{jl} is the corresponding matrix of traces, so the definition of Ric as a bilinear form does not really depend on the choice of orthonormal coordinates for T_pS. Its application to e_1 yields the sum of the sectional curvatures of axis planes containing e_1:

$$e_1 \cdot \text{Ric}(e_1) = R_{11} = \sum_i R_{i1i1} = \sum_{i \neq 1} R_{i1i1}$$

$$= \sum_{i=2}^{m} K(e_1 \wedge e_i).$$

Hence for any orthonormal basis $v_1, \ldots, v_m$ for T_pS,

$$v_1 \cdot \text{Ric}\,(v_1) = \sum_{i=2}^{m} K(v_1 \wedge v_i), \tag{8}$$

and for any unit $v \in T_pS$,

$$v \cdot \text{Ric}\,(v) = \frac{m-1}{\text{vol}\,S^{m-2}} \int_{\substack{w \perp v \\ w \in T_pS}} K(v \wedge w). \tag{9}$$

Thus the Ricci curvature has an interpretation as an average of sectional curvatures.

The *scalar* curvature R is defined as the trace of the Ricci curvature:

$$R = \sum_i R_{ii}. \tag{10}$$

Hence for any orthonormal basis $v_1, \ldots, v_m$ for T_pS,

$$R = 2 \sum_{1 \le i < j \le m} K(v_i \wedge v_j) = \frac{m(m-1)}{\text{vol}\,\mathcal{P}} \int_{P \in \mathcal{P}} K(P) \tag{11}$$

where $\mathcal{P}$ is the set of all 2-planes in T_pS. Thus the scalar curvature is proportional to the average of all sectional curvatures at a point.

Remark.   Historically Ric used to have the opposite sign. Some texts give the Riemannian curvature tensor R_{ijkl} the opposite sign.

5.3. The covariant derivative.   Let S be a C^2 m-dimensional surface in $\mathbf{R}^n$. If f is a differentiable function on S, then the derivative ∇u is a tangent vectorfield. But if $\mathbf{f}$ is a vectorfield (or a field of matrices or tensors), pointwise in T_pS, then the derivative generally will have components normal to S. The projection into T_pS is called the *covariant derivative*. See Figures 5.1 and 5.2. (The name comes from certain nice transformation properties in a more general setting; see Chapter 6.)

In local coordinates $u_1, \ldots, u_m$ in which $g = I$ to first order at p, the coordinates of the covariant derivative of $\mathbf{f}$ at p are given by the several partial derivatives. For example, the coordinates $f_{i;j}$ of the covariant derivative of a vectorfield with coordinates f_i are given by

$$f_{i;j} = f_{i,j} \equiv \frac{\partial f_i}{\partial u_j}.$$

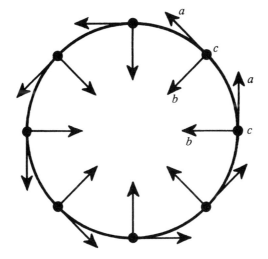

Figure 5.1. (*a*) A vectorfield **f** on the circle, (*b*) its derivative, and (*c*) its covariant derivative, 0.

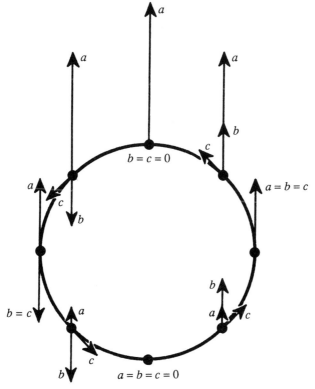

Figure 5.2. (*a*) A vectorfield **f** on the circle, (*b*) its derivative, and (*c*) its covariant derivative.

EXERCISES

5.1. This problem studies the curvature at the origin of the 3-dimensional surface in $\mathbf{R}^5$ given by

$$y_1 = x_1^2 + 2x_1x_2 + x_2^2 + 5x_3^2,$$

$$y_2 = 3x_1^2 + x_2^2 + 2x_2x_3 + x_3^2.$$

  a. What is II (at the origin)?
  b. What is the sectional curvature of the x_1,x_2-plane?
  c. What is the sectional curvature of the plane $x_1 + x_2 = 0$? of the plane $x_1 + x_2 + x_3 = 0$?
  d. Give all the components of the Riemannian curvature tensor. Use them to recompute the answers to parts b and c.
  e. Compute the Ricci and scalar curvatures.

5.2. Consider the vectorfield on $\mathbf{R}^3$: $\mathbf{f} = y^2\mathbf{i} + (x + z)\mathbf{j} + x^3\mathbf{k}$.
  a. Compute its derivative at a general point in $\mathbf{R}^3$.
  b. Compute its covariant derivative at $(0, 0, 1)$ on the unit sphere.

5.3. Show that for the graph of a function $f: \mathbf{R}^{n-1} \to \mathbf{R}$,

$$H = \operatorname{div} \frac{\nabla f}{\sqrt{1 + |\nabla f|^2}} = \frac{(1 + |\nabla f|^2)\Delta f - \Sigma f_i f_j f_{ij}}{(1 + |\nabla f|^2)^{3/2}},$$

where $f_i \equiv \partial f/\partial x_i$, $f_{ij} \equiv \partial^2 f/\partial x_i\,\partial x_j$, $\nabla f \equiv (f_1, \ldots, f_{n-1})$, $\operatorname{div}(p, q, \ldots) \equiv p_1 + q_2 + \cdots$, and $\Delta f \equiv \operatorname{div} \nabla f = f_{11} + f_{22} + \cdots$.

6

Intrinsic Riemannian Geometry

Since many analytic geometric quantities are intrinsic to a smooth m-dimensional surface S in $\mathbf{R}^n$, the standard treatment avoids all reference to an ambient $\mathbf{R}^n$. The surface S is defined as a topological manifold covered by compatible C^∞ coordinate charts, with a "Riemannian metric" g (any smooth positive definite matrix). This is not really a more general setting, since J. Nash [N] has proved that every such abstract Riemannian manifold can be isometrically embedded in some $\mathbf{R}^n$. I suppose that it is a more natural setting, but the formulas get much more complicated.

So far we have seen one intrinsic quantity, the Gaussian curvature G of a 2-dimensional surface in $\mathbf{R}^n$. We proved G intrinsic by deriving a formula for G in terms of the metric.

One may think of intrinsic Riemannian geometry as nothing but a huge collection of such formulas, thus proving intrinsic such quantities as Riemannian curvature, sectional curvature, and covariant derivatives. The standard approach uses these formulas as definitions. We have the advantage of having the simpler extrinsic definitions behind us. Formulas get much more complicated in intrinsic local coordinates.

In particular, complications arise because the local coordinates fail to be orthogonal, as in Figure 6.1. The u_1-axis is not perpendicular to the level set $\{u_1 = 0\}$; or infinitesimally, the unit vector $e_1 = \partial/\partial u_1$ is not perpendicular to the level set $\{du_1 = 0\}$. Hence the

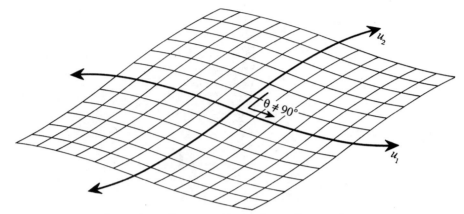

Figure 6.1. For nonorthogonal coordinates, the u_1-axis is not perpendicular to the u_2-axis (the level set $\{u_1 = 0\}$). Infinitesimally, the unit vector $e_1 = \partial/\partial u_1$ is not perpendicular to the level set $\{du_1 = 0\}$.

components of a vectorfield

$$X = (X^1, X^2, \ldots, X^m) = \sum X^i e_i = \sum X^i \frac{\partial}{\partial u^i}$$

and the components of a differential

$$\varphi = (\varphi_1, \varphi_2, \ldots, \varphi_m) = \sum \varphi_i \, du^i$$

behave very differently, under changes in coordinates for example. To emphasize the distinction, superscripts are used on the components of vector-like, or *contravariant*, tensors, and subscripts are used on the components of differential-like, or *covariant*, tensors.

Thus a vectorfield X has components X^i. Its covariant derivative has components $X^i_{;j}$, distinguished by the semicolon from the partial derivatives $X^i_{,j} = \partial X^i / \partial u_j$. As our first exercise in intrinsic Riemannian geometry, we will prove that the components of the covariant derivative are given by the formula

$$X^i_{;j} = X^i_{,j} + \sum_k \Gamma^i_{jk} X^k, \tag{1}$$

where Γ^i_{jk} are the Christoffel symbols

$$\Gamma^i_{jk} = \frac{1}{2} \sum_l g^{il} (g_{lj,k} + g_{lk,j} - g_{jk,l}), \tag{2}$$

defined in terms of the partial derivatives of the metric g_{ij} and its inverse g^{ij}. In particular, covariant differentiation is an intrinsic notion.

In formula (1), the partial derivative gives the first, main term. There are additional terms because the basis vectors themselves are turning.

In both formulas (1) and (2), note how each index i, j, or k on the left occurs in the same position (as a subscript or superscript) on the right. Note how summation runs over the index k or l which appears as both a subscript and a superscript. By these conventions our notation will respect covariance and contravariance.

Some treatments consider covariant differentiation Γ^i_{jk} on manifolds without metrics. Covariant differentiation is also called a *connection*, because by providing for the differentiation of vectorfields it gives some connection between the tangent spaces of S at different points. Our canonical connection which comes from a Riemannian metric (the "Levi-Civita connection") is symmetric, so the *torsion* is 0:

$$T^i_{jk} = \Gamma^i_{jk} - \Gamma^i_{kj} = 0. \tag{3}$$

The Riemannian curvature is given by the formula

$$R^i_{jkl} = -\Gamma^i_{jk,l} + \Gamma^i_{jl,k} + \sum_h (-\Gamma^h_{jk}\Gamma^i_{hl} + \Gamma^h_{jl}\Gamma^i_{hk}). \tag{4}$$

The Riemannian curvature is thus intrinsic, because the connection Γ^i_{jk} is intrinsic. Note that each index on the left occurs in the same position on the right and that summation runs over the index h which appears as both a subscript and a superscript.

The old symmetries 5.2(4–6) still hold for the related tensor

$$R_{ijkl} = \sum_h g_{ih}R^h_{jkl}.$$

Since $R^i_{jkl} = \Sigma_h g^{ih}R_{hjkl}$,

$$R^i_{jlk} = -R^i_{jkl}, \tag{5}$$

but in general $R^j_{ikl} \neq -R^i_{jkl}$. For example, R^2_{2kl} need not vanish. The first Bianchi identity still holds:

$$R^i_{jkl} + R^i_{klj} + R^i_{ljk} = 0. \tag{6}$$

The Ricci curvature is given by the formula

$$R_{jl} = \sum_i R^i_{jil}, \tag{7}$$

and the scalar curvature by the formula

$$R = \sum g^{jl} R_{jl}. \tag{8}$$

The sectional curvature of a plane with orthonormal basis v, w is given by

$$K(v \wedge w) = \sum_{i,j,k,l} R_{ijkl} v^i w^j v^k w^l. \tag{9}$$

If S is 2-dimensional, its Gauss curvature is $G = R/2$.

Note that if $g = I$ to first order at p, then

$$\Gamma^i_{jk} = 0,$$

$$X^i_{;j} = X^i_{,j},$$

$$R_{ijkl} = R^i_{jkl} = -\Gamma^i_{jk,l} + \Gamma^i_{jl,k},$$

$$R_{jl} = \sum_i R^i_{jil},$$

$$R = \sum R_{jj},$$

and

$$K(v \wedge w) = \sum R^i_{jkl} v^i w^j v^k w^l.$$

Remark. An intrinsic definition of the scalar curvature R at a point p in an m-dimensional surface S could be based on the formula for the volume of a ball of intrinsic radius r about p:

$$\text{volume} = \alpha_m r^m - \alpha_m \frac{R}{6(m+2)} r^{m+2} + \cdots, \tag{10}$$

where α_m is the volume of a unit ball in $\mathbf{R}^m$. When $m = 2$, this formula reduces to 3.7(1). The analogous formula for spheres played a role in R. Schoen's solution of the Yamabe problem of finding a conformal deformation of a given Riemannian metric to one of constant scalar curvature [Sc, Lemma 2].

6.1. More useful formulas.   There are a few more special formulas needed sometimes. The covariant derivative of a general tensor f is given by the formula

$$f^{j_1\cdots j_s}_{i_1\cdots i_r;k} = f^{j_1\cdots j_s}_{i_1\cdots i_r,k} + \sum_m \Gamma^{j_1}_{mk} f^{mj_2\cdots j_s}_{i_1\cdots i_r} + \cdots$$

$$+ \sum_m \Gamma^{j_s}_{mk} f^{j_1\cdots j_{s-1}m}_{i_1\cdots i_r} - \sum_m \Gamma^m_{ki_1} f^{j_1\cdots j_s}_{mi_2\cdots i_r} - \cdots \tag{1}$$

$$- \sum_m \Gamma^m_{ki_r} f^{j_1\cdots j_s}_{i_1\cdots i_{r-1}m}.$$

Ricci's lemma says that the covariant derivative of the metric is 0:

$$g_{ij;k} = g^{ij}_{;k} = 0. \tag{2}$$

In general, the mixed partial covariant derivatives of a vector-field X are not equal. *Ricci's identity* gives a very nice formula for the difference in terms of the Riemannian curvature:

$$X^i_{;j;k} - X^i_{;k;j} = - \sum_h R^i_{hjk} X^h. \tag{3}$$

Ricci's identity thus provides an alternative description of the Riemannian curvature as a failure of equality for mixed partials. In intrinsic formulations of Riemannian geometry, Ricci's identity is sometimes turned into a definition of Riemannian curvature.

6.2. Proofs.   There are two ways to prove the intrinsic formulas of Riemannian geometry: either directly from the extrinsic definitions or more intrinsically by exploiting the invariance under changes of coordinates. As an example, we prove the formula for the covariant derivative of a vectorfield both ways and then compare the two methods.

Extrinsic proof of 6.0(1).   Consider a differentiable vectorfield

$$X = \sum_i X^i \frac{\partial}{\partial u^i} = \sum_{k,m} X^k \frac{\partial x^m}{\partial u^k} \frac{\partial}{\partial x^m}.$$

The ordinary partial derivative satisfies

$$\frac{\partial X}{\partial u^j} = \sum_i X^i_{,j} \frac{\partial}{\partial u^i} + \sum_{k,m} X^k \frac{\partial^2 x^m}{\partial u^j \partial u^k} \frac{\partial}{\partial x^m} \tag{1}$$

by the product rule. To obtain the covariant derivative, we project the derivative onto T_pS, the span of $x_1, x_2, \ldots, x_m$, the column space or range of the matrix

$$A = \left[\frac{\partial x^i}{\partial u^j}\right].$$

It is a well-known fact from linear algebra that the projection matrix

$$P = A(A^tA)^{-1}A^t = Ag^{-1}A^t.$$

(A is generally not square, but A^tA is square and invertible.)

In (1), the first summation over i already lies in T_pS. We consider the second summation over k, m. To get the coefficient of $\partial/\partial x^n$ in the projection, we multiply the coefficient of $\partial/\partial x^m$ in (1) by the n,m-entry of $P = Ag^{-1}A^t$, which is

$$\sum_{i,l} \frac{\partial x^n}{\partial u^i} g^{il} \frac{\partial x^m}{\partial u^l},$$

to obtain

$$\sum \frac{\partial x^n}{\partial u^i} g^{il} \frac{\partial x^m}{\partial u^l} \frac{\partial^2 x^m}{\partial u^j \partial u^k} X^k \frac{\partial}{\partial x^n} = \sum_i \left(\sum_l g^{il} \mathbf{x}_{jk} \cdot \mathbf{x}_l\right) X^k \frac{\partial}{\partial u^i}.$$

Therefore

$$X^i_{;j} = X^i_{,j} + \sum_k \Gamma^i_{jk} X^k,$$

where

$$\Gamma^i_{jk} = \sum_l g^{il} \mathbf{x}_{jk} \cdot \mathbf{x}_l$$

$$= \sum_l g^{il} (\tfrac{1}{2})[(\mathbf{x}_l \cdot \mathbf{x}_j)_k + (\mathbf{x}_l \cdot \mathbf{x}_k)_j - (\mathbf{x}_j \cdot \mathbf{x}_k)_l]$$

$$= \tfrac{1}{2} \sum_l g^{il} (g_{lj,k} + g_{lk,j} - g_{jk,l}).$$

Notice how at the final steps we passed from extrinsic quantities $\mathbf{x}_{jk}$ to the intrinsic $g_{rs,t}$.

Invariance proof of 6.0(1). In this method of proof, we first check the formula in coordinates for which $g = I$ to first order at p and then check that the formula is invariant under changes of coordinates.

If $g = I$ to first order, 6.0(1) says that $X^i_{;j} = X^i_{,j}$, which we accept after a few moments' reflection.

To check invariance, we must show that if u^i, $u^{i'}$ give coordinates at p, then

$$X^{i\prime}_{;j} = \frac{\partial u^{i\prime}}{\partial u^m}\frac{\partial u^n}{\partial u^{j\prime}}X^m_{;n},$$

where we henceforth agree to sum over repeated indices. (Getting such formulas right—knowing whether the $\partial u^{i\prime}$ goes on the top or the bottom—is perhaps the hardest part of linear algebra, but our index conventions make it automatic.) This verification is something of a mess, but here we go. First we note that

$$g'_{jl} = \frac{\partial u^r}{\partial u^{j\prime}}\frac{\partial u^s}{\partial u^{l\prime}}g_{rs}.$$

Hence

$$g'_{jl,k} = \frac{\partial u^r}{\partial u^{j\prime}}\frac{\partial u^s}{\partial u^{l\prime}}\frac{\partial u^t}{\partial u^{k\prime}}g_{rs,t} + \left[\frac{\partial^2 u^r}{\partial u^{j\prime}\partial u^{k\prime}}\frac{\partial u^s}{\partial u^{l\prime}} + \frac{\partial u^r}{\partial u^{j\prime}}\frac{\partial^2 u^s}{\partial u^{l\prime}\partial u^{k\prime}}\right]g_{rs},$$

and similar equations hold for $-g_{jk,i}$, $g_{kl,j}$. Combining all three with the definition of Γ^i_{jk} and

$$g^{il\prime} = \frac{\partial u^{i\prime}}{\partial u^p}\frac{\partial u^{l\prime}}{\partial u^q}g^{pq}$$

yields

$$\Gamma^{i\prime}_{jk} = \frac{1}{2}\frac{\partial u^{i\prime}}{\partial u^p}\frac{\partial u^{l\prime}}{\partial u^q}g^{pq}\left(\frac{\partial u^r}{\partial u^{j\prime}}\frac{\partial u^s}{\partial u^{l\prime}}\frac{\partial u^t}{\partial u^{k\prime}}[g_{rs,t} - g_{rt,s} + g_{ts,r}] + \Omega\right),$$

where

$$\Omega = g_{rs}\left[\frac{\partial^2 u^r}{\partial u^{j\prime}\partial u^{k\prime}}\frac{\partial u^s}{\partial u^{l\prime}} + \frac{\partial u^r}{\partial u^{j\prime}}\frac{\partial^2 u^s}{\partial u^{l\prime}\partial u^{k\prime}} - \frac{\partial^2 u^r}{\partial u^{j\prime}\partial u^{l\prime}}\frac{\partial u^s}{\partial u^{k\prime}}\right.$$
$$\left. - \frac{\partial u^r}{\partial u^{j\prime}}\frac{\partial^2 u^s}{\partial u^{k\prime}\partial u^{l\prime}} + \frac{\partial^2 u^r}{\partial u^{k\prime}\partial u^{j\prime}}\frac{\partial u^s}{\partial u^{l\prime}} + \frac{\partial u^r}{\partial u^{k\prime}}\frac{\partial^2 u^s}{\partial u^{j\prime}\partial u^{l\prime}}\right]$$
$$= 2g_{rs}\left[\frac{\partial^2 u^r}{\partial u^{j\prime}\partial u^{k\prime}}\frac{\partial u^s}{\partial u^{l\prime}}\right],$$

because g_{rs} is symmetric.

Therefore

$$\Gamma^{i\prime}_{jk} = \frac{1}{2}\frac{\partial u^{i\prime}}{\partial u^p}\frac{\partial u^r}{\partial u^{j\prime}}\frac{\partial u^t}{\partial u^{k\prime}}\delta^s_q g^{pq}(g_{rs,t} - g_{rt,s} + g_{ts,r})$$

$$+ \frac{\partial^2 u^r}{\partial u^{j\prime}\partial u^{k\prime}}\frac{\partial u^{i\prime}}{\partial u^p}\delta^s_q g^{pq}g_{rs},$$

where $\delta^s_q = 1$ if $s = q$ and 0 otherwise. Since $\delta^s_q g^{pq}g_{rs} = g^{pq}g_{qr} = \delta^p_r$,

$$\Gamma^{i\prime}_{jk} = \frac{\partial u^{i\prime}}{\partial u^p}\frac{\partial u^r}{\partial u^{j\prime}}\frac{\partial u^t}{\partial u^{k\prime}}\left[\frac{1}{2}g^{ps}(g_{rs,t} - g_{rt,s} + g_{ts,r})\right] + \frac{\partial^2 u^r}{\partial u^{j\prime}\partial u^{k\prime}}\frac{\partial u^{i\prime}}{\partial u^r}$$

$$= \frac{\partial u^{i\prime}}{\partial u^p}\frac{\partial u^r}{\partial u^{j\prime}}\frac{\partial u^t}{\partial u^{k\prime}}\Gamma^p_{rt} + \frac{\partial^2 u^h}{\partial u^{j\prime}\partial u^{k\prime}}\frac{\partial u^{i\prime}}{\partial u^h},$$

by changing the dummy index in the last term from r to h.

Multiplying both sides by $\partial u^h/\partial u^{i\prime}$ and changing primes and indices yields

$$\frac{\partial^2 u^{i\prime}}{\partial u^m \partial u^l} = \Gamma^h_{lm}\frac{\partial u^{i\prime}}{\partial u^h} = \Gamma^{i\prime}_{hk}\frac{\partial u^{h\prime}}{\partial u^l}\frac{\partial u^{k\prime}}{\partial u^m}. \qquad (2)$$

Now

$$X^{i\prime}_{;j} \equiv \frac{\partial}{\partial u^{j\prime}}X^{i\prime} + \Gamma^{i\prime}_{jk}X^{k\prime}$$

$$= \frac{\partial}{\partial u^{j\prime}}\left(\frac{\partial u^{i\prime}}{\partial u^m}X^m\right) + \Gamma^{i\prime}_{jk}\left(\frac{\partial u^{k\prime}}{\partial u^m}X^m\right)$$

$$= X^m_{,n}\frac{\partial u^n}{\partial u^{j\prime}}\frac{\partial u^{i\prime}}{\partial u^m} + X^m\frac{\partial^2 u^{i\prime}}{\partial u^m \partial u^l}\frac{\partial u^l}{\partial u^{j\prime}} + X^m\frac{\partial u^{k\prime}}{\partial u^m}\Gamma^{i\prime}_{jk}.$$

By (2),

$$X^{i\prime}_{;j} = X^m_{,n}\frac{\partial u^n}{\partial u^{j\prime}}\frac{\partial u^{i\prime}}{\partial u^m}$$

$$+ X^m\left[\Gamma^h_{lm}\frac{\partial u^{i\prime}}{\partial u^h}\frac{\partial u^l}{\partial u^{j\prime}} - \Gamma^{i\prime}_{hk}\frac{\partial u^{h\prime}}{\partial u^l}\frac{\partial u^{k\prime}}{\partial u^m}\frac{\partial u^l}{\partial u^{j\prime}} + \Gamma^{i\prime}_{jk}\frac{\partial u^{k\prime}}{\partial u^m}\right]$$

$$= \frac{\partial u^{i\prime}}{\partial u^m}\frac{\partial u^n}{\partial u^{j\prime}}X^m_{,n} + \frac{\partial u^{i\prime}}{\partial u^h}\frac{\partial u^l}{\partial u^{j\prime}}\Gamma^h_{lm}X^m.$$

By changing dummy variables in the second term $(m \to k,$
$h \to m, l \to n)$, we obtain

$$X^i_{;j} = \frac{\partial u^{i'}}{\partial u^m}\frac{\partial u^n}{\partial u^{j'}}[X^m_{,n} + \Gamma^m_{nk}X^k] = \frac{\partial u^{i'}}{\partial u^m}\frac{\partial u^n}{\partial u^{j'}}X^m_{;n},$$

as desired.

Remark. Of the two proofs, the first has the advantages of being shorter and deriving the formula, whereas the second proves a given formula.

6.3. Geodesics. Let C be a C^2 curve in a C^2 m-dimensional surface S in $\mathbf{R}^n$, with curvature vector $\boldsymbol{\kappa}$ at a point $p \in C$. We define the *geodesic curvature* $\boldsymbol{\kappa}_g$ as the projection of $\boldsymbol{\kappa}$ onto the tangent space T_pS. Equivalently, $\boldsymbol{\kappa}_g$ is the covariant derivative of the unit tangent vector. While curvature $\boldsymbol{\kappa}$ is extrinsic, geodesic curvature $\boldsymbol{\kappa}_g$ is intrinsic.

A *geodesic* is a curve with $\boldsymbol{\kappa}_g = \mathbf{0}$ at all points. For example, geodesics on spheres are arcs of great circles, but other circles of latitude are not geodesics. (See Figure 6.2.) Shortest paths turn out to be geodesics, but there are sometimes also other longer geodesics between pairs of points. For example, nonantipodal points on the equator are joined by a short and a long geodesic, depending on

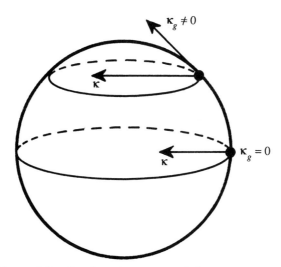

Figure 6.2. On the sphere, great circles are geodesics ($\boldsymbol{\kappa}_g = \mathbf{0}$), but other circles of latitude are not ($\boldsymbol{\kappa}_g \neq \mathbf{0}$).

which way you go. The poles are joined by infinitely many semicircular meridians of longitude, all of the same length.

The following theorem explains why shortest paths must be geodesics.

6.4. Theorem.   *A curve is a geodesic if and only if the first variation of its length vanishes.*

Proof.   Let $\mathbf{x}(t)$ be a local parametrization by arc length. Corresponding to an infinitesimal, compactly supported change $\delta\mathbf{x}$ in $\mathbf{x}(t)$ is a variation in length

$$\delta L = \delta \int (\dot{\mathbf{x}} \cdot \dot{\mathbf{x}})^{1/2} = \int \tfrac{1}{2}(\dot{\mathbf{x}} \cdot \dot{\mathbf{x}})^{-1/2} 2\dot{\mathbf{x}} \cdot \delta\dot{\mathbf{x}}$$

$$= \int \mathbf{T} \cdot \delta\dot{\mathbf{x}} = - \int \dot{\mathbf{T}} \cdot \delta\mathbf{x} = - \int \boldsymbol{\kappa} \cdot \delta\mathbf{x} = - \int \boldsymbol{\kappa}_g \cdot \delta\mathbf{x}$$

by integration by parts and the fact that $\delta\mathbf{x}$ stays in the surface. δL vanishes for all $\delta\mathbf{x}$ along the surface if and only if $\boldsymbol{\kappa}_g = \mathbf{0}$ and the curve is a geodesic.

Remark.   It follows from the theory of differential equations that in any C^2 m-dimensional surface, through any point in any direction there is locally a unique C^2 geodesic. Such a geodesic provides the shortest path to nearby points. A little more argument shows that if S is connected and compact (or connected and merely complete), between *any* two points some geodesic provides the shortest path (the Hopf-Rinow Theorem, see [CE, ch. 3] or [He, Theorem 10.4]).

6.5. Formula for geodesics.   In local coordinates $u^1, \ldots, u^m$, consider a curve $u(t)$ parametrized by arc length so that the unit tangent vector $\mathbf{T} = \dot{u}$. The derivative of any function $f(u)$ along the curve is given by $\Sigma f_{,j}\dot{u}^j$ (the chain rule). The covariant derivative of any vectorfield X^i along the curve satisfies

$$\sum_j X^i_{;j}\dot{u}^j = \sum_j X^i_{,j}\dot{u}^j + \sum_{j,k} \Gamma^i_{jk}\dot{u}^j X^k \qquad (1)$$

$$= \dot{X}^i + \sum_{j,k} \Gamma^i_{jk}\dot{u}^j X^k$$

[see 6.0(1)]. Hence for a geodesic (parametrized by arc length), the covariant derivative along the curve of the vectorfield $X^i = \mathbf{T}^i = \dot{u}^i$ must vanish:

$$0 = \ddot{u}^i + \sum_{j,k} \Gamma^i_{jk} \dot{u}^j \dot{u}^k. \tag{2}$$

6.6. Hyperbolic geometry.   As an example in Riemannian geometry, we consider 2-dimensional hyperbolic space H for which global coordinates are given by the upper halfplane

$$\{(x, y) \in \mathbf{R}^2 : y > 0\}$$

with metric

$$g_{ij} = y^{-2} \delta_{ij};$$

that is,

$$ds^2 = \frac{1}{y^2} dx^2 + \frac{1}{y^2} dy^2.$$

Since pointwise g is a multiple of the standard metric (g is "conformal"), angles are the same in the upper halfplane as on H, although distances are different, of course.

Now we compute the Christoffel symbols and curvature.

$$g^{ij} = y^2 \delta^{ij}.$$

By formula 6.0(2),

$$\Gamma^1_{12} \equiv \Gamma^1_{21} = \frac{1}{2} y^2 (g_{12,1} + g_{11,2} - g_{12,1})$$

$$= \frac{1}{2} y^2 \frac{\partial}{\partial y} y^{-2} = -y^{-1}.$$

Similarly,

$$-\Gamma^2_{11} = \Gamma^2_{22} = -y^{-1}$$

and the rest are 0.

By formula 6.0(4),

$$R^1_{212} = -\Gamma^1_{21,2} + \Gamma^1_{22,1} + \sum_h (-\Gamma^h_{21}\Gamma^1_{h2} + \Gamma^h_{22}\Gamma^1_{h1})$$

$$= -y^{-2} + 0 + (-y^{-2} + y^{-2}) = -y^{-2}.$$

Similarly,

$$R^2_{121} = -y^{-2}, \qquad R^1_{111} = R^2_{222} = 0,$$

$$R_{11} = R^1_{111} + R^2_{121} = -y^{-2}, \qquad R_{22} = -y^{-2},$$

$$R = -2, \qquad G = -1.$$

Thus hyperbolic space H has constant Gaussian curvature -1 and assumes its exalted place with the plane ($G = 0$) and the sphere ($G = 1$).

Geodesics parametrized by arc length t must satisfy the equations 6.5(2):

$$\ddot{x} - 2y^{-1}\dot{x}\dot{y} = 0 \qquad \text{and} \qquad \ddot{y} + y^{-1}\dot{x}^2 - y^{-1}\dot{y}^2 = 0.$$

Let $p = dx/dy$; then

$$\dot{x} = p\dot{y}, \qquad \ddot{x} = \frac{dp}{dy}\dot{y}^2 + p\ddot{y}.$$

Substituting for $\ddot{y}$ from the second equation in the first yields

$$\frac{dp}{dy} = y^{-1}(p^3 + p).$$

Integrating by partial fractions gives

$$\frac{dx}{dy} = p = \pm\frac{cy}{\sqrt{1 - c^2y^2}}.$$

If $c = 0$, we obtain vertical lines as geodesics. Otherwise, letting $c = 1/a$ and integrating yields

$$(x - b)^2 + y^2 = a^2.$$

These geodesics are just semicircles centered on the x-axis. See Figure 6.3.

Through any two points there is a unique geodesic, or "straight line," that provides the shortest path between the points. Indeed, Euclid's first four postulates all hold. The notorious fifth postulate fails. Its equivalent statement due to Playfair says that for a given line l and a point p not on the line, there is a unique line through p that does not intersect l. The uniqueness fails in hyperbolic geometry, as Figure 6.3 illustrates. Thus hyperbolic geometry proves the impossibility of what geometers had been attempting for millenia—to deduce the fifth postulate from the first four—and gives the premier example of non-Euclidean geometry.

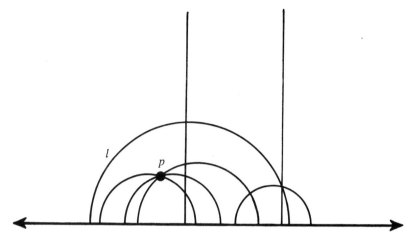

Figure 6.3.   Geodesics in hyperbolic space H are semicircles centered on the x-axis and vertical lines.

It is interesting to note that the hyperbolic distance from any point (a, b) to the x-axis, measured along a vertical geodesic, is

$$\int_{y=0}^{b} y^{-1}\, dy = \infty.$$

Hyperbolic space actually has no boundary, but extends infinitely far in all directions.

6.7. Geodesics and sectional curvature.   We remark that positive sectional curvature means that parallel geodesics converge, as on the sphere. Negative sectional curvature means that parallel geodesics diverge, as in hyperbolic space.

EXERCISES

6.1. *A torus.* Let T be the torus obtained by revolving a unit circle parametrized by $0 \leq \varphi < 2\pi$ about an axis 4 units from its center. Use the angle $0 \leq \theta < 2\pi$ of revolution and φ as coordinates. (See Figure 6.4.)

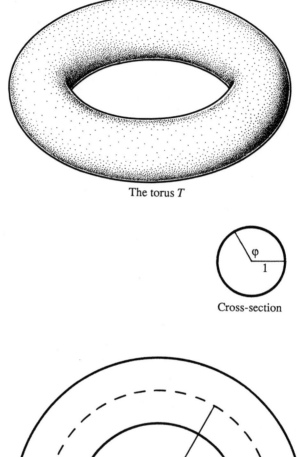

The torus T

Cross-section

Top view

Figure 6.4. The torus T, with coordinates θ, φ.

a. Show that

$$g_{11} = (4 + \cos \varphi)^2, \qquad g_{12} = g_{21} = 0, \qquad g_{22} = 1,$$

$$\Gamma^1_{12} = -\frac{\sin \varphi}{4 + \cos \varphi}, \qquad \Gamma^2_{11} = \sin \varphi(4 + \cos \varphi),$$

the rest are 0.

b. Consider the vectorfield

$$a = \cos \varphi \frac{\partial}{\partial \theta},$$

that is, $a^1 = \cos \varphi$ and $a^2 = 0$. Show that the covariant derivative is given by

$$a^1_{;2} = -\sin \varphi \frac{4 + 2\cos \varphi}{4 + \cos \varphi}, \qquad a^2_{;1} = \sin \varphi \cos \varphi(4 + \cos \varphi),$$

the rest are 0.

c. Show that the length of the spiral curve $\theta(t) = \varphi(t) = t$ $(0 \le t \le 2\pi)$ is given by the integral

$$\int_0^{2\pi} [(4 + \cos t)^2 + 1]^{1/2} \, dt.$$

d. Of course, $R^1_{111} = R^2_{222} = 0$. Show that

$$R^1_{212} = \frac{\cos \varphi}{4 + \cos \varphi}, \qquad R^2_{121} = \cos \varphi(4 + \cos \varphi),$$

$$R_{11} = \cos \varphi(4 + \cos \varphi), \qquad R_{12} \equiv R_{21} = 0,$$

$$R_{22} = \frac{\cos \varphi}{4 + \cos \varphi}, \qquad R = \frac{2\cos \varphi}{4 + \cos \varphi}, \qquad G = \frac{\cos \varphi}{4 + \cos \varphi}.$$

e. Conclude that there is no distance-preserving map of any region in the torus on regions in the plane or on the sphere.

6.2. *The sphere.* This exercise will verify that geodesics are arcs of great circles and that the sphere has constant Gauss curvature.

Consider a sphere S of radius a with the usual spherical coordinates $u^1 = \theta$ and $u^2 = \varphi$.

a. Check that a great circle is given by the equation

$$w \equiv \cot \varphi = c_1 \cos \theta + c_2 \sin \theta,$$

except for vertical great circles $\theta = c$. Hint: A nonvertical great circle is the intersection of a sphere with a plane

$$z = c_1 x + c_2 y.$$

b. Show that the metric is given by

$$g_{11} = a^2 \sin^2 \varphi, \qquad g_{12} = 0, \qquad g_{22} = a^2.$$

c. Compute the Christoffel symbols

$$\Gamma^1_{12} = \cot \varphi, \qquad \Gamma^2_{11} = - \sin \varphi \cos \varphi,$$

the rest vanish. Conclude that geodesics satisfy

$$\begin{cases} \ddot\theta + 2 \cot \varphi \quad \dot\theta\dot\varphi = 0 \\ \ddot\varphi - \sin \varphi \cos \varphi \ \dot\theta^2 = 0 \end{cases}$$

and hence $\varphi'' - 2 \cot \varphi \, \varphi'^2 - \sin \varphi \cos \varphi = 0$, where primes denote derivatives with respect to θ, unless $\theta = c$. Thus $w'' + w = 0$, so

$$w = c_1 \cos \theta + c_2 \sin \theta.$$

Therefore geodesics are indeed arcs of great circles.

d. Compute the Riemannian curvature

$$R^1_{212} = - R^1_{221} = 1, \qquad R^2_{121} = - R^2_{112} = \sin^2 \varphi,$$

the rest are 0; the Ricci curvature

$$R_{11} = \sin^2 \varphi, \qquad R_{12} = R_{21} = 0, \qquad R_{22} = 1;$$

the scalar curvature $R = 2a^{-2}$; and finally the Gauss curvature $G = a^{-2}$.

Remark. Actually a simple symmetry argument shows that geodesics are arcs of great circles. We may assume that the geodesic is tangent to the equator at a point. By uniqueness, it must equal its own reflection across the equator. Hence it must be an arc of the equator.

7

General Relativity

Toward the end of the nineteenth century, a puzzling inconsistency in Mercury's orbit was observed.

Newton had brilliantly explained Kepler's elliptical planetary orbits through solar gravitational attraction and calculus. His successors used a method of perturbations to compute the deviations caused by the other planets. Their calculations predicted that the elliptical orbit shape should rotate, or precess, some fraction of a degree per century:

Planet	Predicted Precession (per century)
Saturn	46'
Jupiter	432"
Mercury	532"

Here 60' (60 minutes) equals 1 degree of arc, and 60" (60 seconds) equals 1 minute of arc.

Observation confirmed the predictions for Saturn and Jupiter, but showed that Mercury precessed 575" per century. By 1900, it was obvious that the variance from the expected precession exceeded any conceivable experimental error. What was causing the additional 43" per century?

General relativity would provide the answer.

Figure 7.1. "Mercury's running slow."

7.1. General relativity. The theory of general relativity has three elements. First, special relativity describes motion in free space without gravity. Second, the Principle of Equivalence extends the theory, at least in principle, to include gravity, roughly by equating gravity with acceleration. Third, Riemannian geometry provides a mathematical framework which makes calculations possible.

I first learned the derivation of Mercury's precession from Spain [Sp, Chapter VIII] and Weinberg [W, Chapter 9] with the help of my friend Ira Wasserman. The short derivation given here is based on a talk by my student Phat Vu at a mathematics colloquium at Williams College, in turn based on Jeffery [Je, Chap. VII]. A simplified account including some dramatic episodes from the history of astronomy appears in [M4].

7.2. Special relativity. A single particle in free space follows a straight line at constant velocity. For example, $x = at$, $y = bt$, $z = ct$, or

$$\frac{x}{a} = \frac{y}{b} = \frac{z}{c},$$

which is the formula for a straight line through the origin in 3-space. This path is also a straight line in 4-dimensional space-time:

$$\frac{x}{a} = \frac{y}{b} = \frac{z}{c} = \frac{t}{1},$$

that is, a geodesic for the standard metric

$$ds^2 = dx^2 + dy^2 + dz^2 + dt^2. \tag{1}$$

It is actually a geodesic for any metric of the form

$$ds^2 = a_1\, dx^2 + a_2\, dy^2 + a_3\, dz^2 + a_4\, dt^2. \tag{2}$$

Einstein based special relativity on two axioms:

1. The laws of physics look the same in all inertial frames of reference—that is, to all observers moving with constant velocity relative to one another. (Of course, in accelerating reference frames, the laws of physics look different. Cups of lemonade in accelerating cars suddenly fall over, and tennis balls on the floors of rockets flatten like pancakes.)
2. The speed of a light beam is the same relative to any inertial frame, whether moving in the same or the opposite direction. (Einstein apparently guessed this surprising fact without knowing the evidence provided by the famous Michelson-Morley experiment. It leads to other curiosities, such as time's slowing down at high velocities.)

Einstein's postulates hold for motion along geodesics in space-time if one takes the special case of (2):

$$ds^2 = -dx^2 - dy^2 - dz^2 + c^2\, dt^2. \tag{3}$$

This is the famous Lorentz metric, with c the speed of light. We will choose units to make $c = 1$.

The Lorentz metric remains invariant under inertial changes of coordinates, but looks funny in accelerating coordinate systems.

For us a new feature of this metric is the presence of minus signs; the metric is not positive definite. Except for the novel fact that the square of the length of curves in space-time can be positive or negative, all definitions and properties remain formally the same. In particular, positive sectional curvature means that parallel geodesics converge (the square of the distance between them decreases). (See Section 6.7.)

This new distance s is often called "proper time" τ, since a motionless particle (x, y, z constant) has $ds^2 = dt^2$. If we replace the symbol s by τ and change to spherical coordinates, the Lorentz metric becomes

$$d\tau^2 = -dr^2 - r^2\, d\varphi^2 - r^2 \sin^2 \varphi\, d\theta^2 + dt^2. \qquad (4)$$

7.3. The Principle of Equivalence.   Special relativity handles motion — position, velocity, acceleration — in free space. The remaining question is how to handle gravity. The Principle of Equivalence asserts that infinitesimally the physical effects of gravity are indistinguishable from those of acceleration. If you feel pressed against the floor of a tiny elevator, you cannot tell whether it is because the elevator is resting on a massive planet or because the elevator is accelerating upward. Consequently, the effect of gravity is just like that of acceleration: it just makes the formula for ds^2 look funny. Computing motion in a gravitational field will reduce to computing geodesics in some strange metric.

7.4. The Schwarzschild metric.   The most basic example in general relativity is the effect on the Lorentz metric of a single point mass, such as a sun at the center of a solar system. We will assume that the metric takes the simple form

$$d\tau^2 = -e^{\lambda(r)}\, dr^2 - r^2(d\varphi^2 + \sin^2 \varphi\, d\theta^2) + e^{\nu(r)}\, dt^2, \qquad (1)$$

where $\lambda(r)$ and $\nu(r)$ are functions to be determined. This metric is spherically symmetric and time-independent. For physical reasons, Einstein further assumed that what is now called the Einstein tensor vanishes:

$$G^i_k = g^{ij}R_{jk} - \tfrac{1}{2}R\delta^i_k = 0. \qquad (2)$$

To employ this assumption, we now compute the Einstein tensor for the metric (1). We order the variables r, φ, θ, t. We compute first the metric

$$g_{11} = -e^{\lambda}, \qquad g_{22} = -r^2, \qquad g_{33} = -r^2 \sin^2 \varphi, \qquad g_{44} = e^{\nu},$$

others vanish,

$$g^{11} = -e^{-\lambda}, \qquad g^{22} = -r^{-2}, \qquad g^{33} = -r^{-2} \sin^{-2} \varphi,$$
$$g^{44} = e^{-\nu}, \qquad \text{others vanish;}$$

then the Christoffel symbols

$$\Gamma^1_{11} = \tfrac{1}{2}\lambda', \qquad \Gamma^1_{22} = -re^{-\lambda}, \qquad \Gamma^1_{33} = -re^{-\lambda} \sin^2 \varphi, \qquad (3)$$
$$\Gamma^1_{44} = \tfrac{1}{2}\nu'e^{\nu-\lambda}, \qquad \Gamma^2_{12}\Gamma^3_{13} = r^{-1}, \qquad \Gamma^2_{33} = -\sin \varphi \cos \varphi,$$
$$\Gamma^3_{23} = \cot \varphi, \qquad \Gamma^4_{14} = \tfrac{1}{2}\nu', \qquad \text{others vanish,}$$

where λ' denotes $d\lambda/dr$; then some components of the Riemannian curvature tensor

$$R^2_{121} = R^3_{131} = \tfrac{1}{2}r^{-1}\lambda', \qquad R^4_{141} = -\tfrac{1}{2}\nu'' + (\tfrac{1}{2}\nu')(\tfrac{1}{2}\lambda') - \tfrac{1}{4}\nu'^2,$$
$$R^1_{212} = \tfrac{1}{2}r\lambda'e^{-\lambda}, \qquad R^3_{232} = 1 - e^{-\lambda}, \qquad R^4_{242} = \tfrac{1}{2}\nu'(-re^{-\lambda}),$$
$$R^1_{313} = \tfrac{1}{2}r\lambda' \sin^2 \varphi\, e^{-\lambda}, \qquad R^2_{323} = \sin^2 \varphi\,(1 - e^{-\lambda}),$$
$$R^4_{343} - \tfrac{1}{2}\nu'(-r)e^{-\lambda} \sin^2 \varphi,$$
$$R^1_{414} = \tfrac{1}{2}e^{\nu-\lambda}(\nu'' + \tfrac{1}{2}\nu'^2 - \tfrac{1}{2}\nu'\lambda'),$$
$$R^2_{424} = R^3_{434} = \tfrac{1}{2}r^{-1}\nu'e^{\nu-\lambda};$$

then some components of the Ricci curvature

$$R_{11} = r^{-1}\lambda' - \tfrac{1}{2}\nu'' + \tfrac{1}{4}\nu'\lambda' - \tfrac{1}{4}\nu'^2,$$
$$R_{22} = 1 + \tfrac{1}{2}re^{-\lambda}(\lambda' - \nu') - e^{-\lambda},$$
$$R_{33} = \sin^2 \varphi\, R_{22},$$
$$R_{44} = \tfrac{1}{2}e^{\nu-\lambda}(\nu'' + \tfrac{1}{2}\nu'^2 - \tfrac{1}{2}\nu'\lambda' + 2r^{-1}\nu'),$$
$$R = -2r^{-2} + e^{-\lambda}(\nu'' - 2r^{-1}\lambda' - \tfrac{1}{2}\nu'\lambda' + \tfrac{1}{2}\nu'^2 + 2r^{-1}\nu' + 2r^{-2});$$

and finally some components of the Einstein tensor

$$G^i_k = g^{ij}R_{jk} - \tfrac{1}{2}R\delta^i_k,$$
$$G^1_1 = r^{-2} + e^{-\lambda}(-r^{-1}\nu' - r^{-2}),$$
$$G^2_2 = G^3_3 = e^{-\lambda}(-\tfrac{1}{2}\nu'' + \tfrac{1}{2}r^{-1}\lambda' - \tfrac{1}{2}r^{-1}\nu' + \tfrac{1}{4}\nu'\lambda' - \tfrac{1}{4}\nu'^2),$$
$$G^4_4 = r^{-2} + e^{-\lambda}(r^{-1}\lambda' - r^{-2}).$$

Since $G^4_4 = 0$, $e^{-\lambda} = 1 - \gamma r^{-1}$ for some constant γ (just check that

$d\gamma/dr = 0$). Consideration of a test particle with 0 velocity and large r (see Exercise 7.1) leads to the conclusion that $\gamma = 2GM$, where M is the central mass and G is the gravitational constant. Therefore

$$e^{-\lambda} = 1 - 2GMr^{-1}.$$

Since $G_1^1 = G_4^4 = 0$, $\lambda + \nu$ is constant. Since the metric should look like the Lorentz metric for r huge, we conclude that $\lambda + \nu = 0$. Therefore

$$e^{\nu} = e^{-\lambda} = 1 - 2GMr^{-1}. \tag{4}$$

Now $G_2^2 = G_3^3 = 0$ automatically.

We have obtained the famous Schwarzschild metric

$$d\tau^2 = -(1 - 2GMr^{-1})^{-1} dr^2 - r^2(d\varphi^2 + \sin^2 \varphi \, d\theta^2)$$
$$+ (1 - 2GMr^{-1}) dt^2. \tag{5}$$

Notice that if $M = 0$, the Schwarzschild metric (5) reduces to the Lorentz metric 7.2(4). Notice too that as r decreases to $1/2GM$, $d\tau^2$ blows up: shrinking the sun to a point mass has created a black hole of radius $1/2GM$!

7.5. Relativistic celestial mechanics.   Now we are ready to see what differences general relativity predicts for Mercury's orbit. The physics is embodied in the four equations for geodesics 6.5(2) in the Schwarzschild metric 7.4(5). Four equations should let us solve for r, φ, θ, and t as functions of τ. Actually, instead of the first equation for geodesics involving $d^2r/d\tau^2$, we will use the identity $d\tau^2 = g_{ij} \, dx^i \, dx^j$:

$$-(1 - 2GMr^{-1})^{-1}\left(\frac{dr}{d\tau}\right)^2 - r^2\left(\frac{d\varphi}{d\tau}\right)^2 \tag{I}$$

$$- r^2 \sin^2 \varphi \left(\frac{d\theta}{d\tau}\right)^2 + (1 - 2GMr^{-1})\left(\frac{dt}{d\tau}\right)^2 = 1.$$

To compute the three other geodesic equations, we proceed from 7.4(4) to compute

$$\lambda' = -\nu' = -2GM(r^2 - 2GMr)^{-1},$$
$$\lambda'' = -\nu'' = 2GM(r^2 - 2GMr)^{-2}(2r - 2GM),$$

and then, from 7.4(3),

$$\Gamma_{11}^1 = -GM(r^2 - 2GMr)^{-1}$$

$$\Gamma^1_{22} = -r(1 - 2GMr^{-1}) = 2GM - r$$

$$\Gamma^1_{33} = (2GM - r) \sin^2 \varphi$$

$$\Gamma^1_{44} = \tfrac{1}{2}(1 - 2GMr^{-1})(2GMr^{-2})$$

$$\Gamma^2_{12} = \Gamma^3_{13} = r^{-1}$$

$$\Gamma^2_{33} = -\sin \varphi \cos \varphi$$

$$\Gamma^3_{23} = \cot \varphi$$

$$\Gamma^4_{14} = GM(r^2 - 2GMr)^{-1}.$$

Hence the last three geodesic equations [compare to 6.5(2)] are

$$\frac{d^2\varphi}{d\tau^2} + 2r^{-1}\frac{dr}{d\tau}\frac{d\varphi}{d\tau} - \sin \varphi \cos \varphi \left(\frac{d\theta}{d\tau}\right)^2 = 0, \qquad \text{(II)}$$

$$\frac{d^2\theta}{d\tau^2} + 2r^{-1}\frac{dr}{d\tau}\frac{d\theta}{d\tau} + 2 \cot \varphi \frac{d\varphi}{d\tau}\frac{d\theta}{d\tau} = 0, \qquad \text{(III)}$$

$$\frac{d^2t}{d\tau^2} + \frac{2GM}{r^2 - 2GMr}\frac{dr}{d\tau}\frac{dt}{d\tau} = 0. \qquad \text{(IV)}$$

The solution of equations I through IV will give Mercury's orbit. Assuming that initially $d\varphi/d\tau$ and $\cos \varphi$ are 0, by (II) φ remains $\pi/2$. Thus, even relativistically, the orbit remains planar. The other three equations become

$$-(1 - 2GMr^{-1})^{-1}\left(\frac{dr}{d\tau}\right)^2 - r^2\left(\frac{d\theta}{d\tau}\right)^2 + (1 - 2GMr^{-1})\left(\frac{dt}{d\tau}\right)^2 = 1, \qquad \text{(I$'$)}$$

$$\frac{d^2\theta}{d\tau^2} + 2r^{-1}\frac{dr}{d\tau}\frac{d\theta}{d\tau} = 0, \qquad \text{(III$'$)}$$

$$\frac{d^2t}{d\tau^2} + 2GM(r^2 - 2GMr)^{-1}\frac{dr}{d\tau}\frac{dt}{d\tau} = 0. \qquad \text{(IV$'$)}$$

Integrating III$'$ and IV$'$ yields

$$r^2\frac{d\theta}{d\tau} = h \qquad \text{(a constant)}, \qquad \text{(III$''$)}$$

$$(1 - 2GMr^{-1})\frac{dt}{d\tau} = \beta \qquad \text{(a constant)}. \qquad \text{(IV$''$)}$$

Therefore I' becomes

$$-r^{-4}\left(\frac{dr}{d\theta}\right)^2 - r^{-2}(1 - 2GMr^{-1}) + \beta^2 h^{-2} = h^{-2}(1 - 2GMr^{-1}). \quad \text{(I'')}$$

Putting $r = u^{-1}$ yields

$$\left(\frac{du}{d\theta}\right)^2 = 2GM\left(u^3 - \frac{1}{2GM}u^2 + \beta_1 u + \beta_0\right)$$

for some constants β_0, β_1. The maximum and minimum values u_1, u_2 of u must be roots. Since the roots sum to $1/2GM$, the third root is $1/2GM - u_1 - u_2$, and hence

$$\left(\frac{du}{d\theta}\right)^2 = 2GM(u - u_1)(u - u_2)\left(u - \frac{1}{2GM} + u_1 + u_2\right),$$

$$\frac{d\theta}{|du|} = \frac{1}{\sqrt{(u_1 - u)(u - u_2)}}[1 - 2GM(u + u_1 + u_2)]^{-1/2}$$

$$\approx \frac{1 + GM(u + u_1 + u_2)}{\sqrt{(u_1 - u)(u - u_2)}}.$$

To first approximation the orbit is the classical ellipse

$$u = l^{-1}(1 + e \cos \theta),$$

with $u_1 = l^{-1}(1 + e)$, $u_2 = l^{-1}(1 - e)$, and mean distance

$$a = \frac{1}{2}\left(\frac{1}{u_1} + \frac{1}{u_2}\right) = \frac{l}{1 - e^2}.$$

For one revolution,

$$\Delta\theta \approx \int_{\theta=0}^{2\pi} \frac{1 + GMl^{-1}(3 + e \cos \theta)}{\sqrt{l^{-1}e(1 - \cos \theta)l^{-1}e(1 + \cos \theta)}} l^{-1} e|\sin \theta| \, d\theta$$

$$= \int_{\theta=0}^{2\pi} 1 + GMl^{-1}(3 + e \cos \theta) \, d\theta$$

$$= 2\pi + 6\pi GM/l$$

$$= 2\pi + 6\pi GM/a(1 - e^2).$$

The ellipse has precessed $6\pi GM/a(1 - e^2)$ radians. The rate of precession in terms of Mercury's period T is

$$\frac{6\pi GM}{a(1 - e^2)T},$$

or, back in more standard units (in which the speed of light c is not 1),

$$\frac{6\pi GM}{c^2 a(1 - e^2)T} \quad \text{radians.}$$

Now

G = gravitational constant = 6.67×10^{-11} m^3/kg sec^2,

M = mass of sun = 1.99×10^{30} kg,

c = speed of light = 3.00×10^8 m/sec,

a = mean distance from Mercury to sun = 5.768×10^{10} m,

e = eccentricity of Mercury's orbit = 0.206,

T = period of Mercury = 88.0 days,

century = 36525 days,

radian $-$ $360/2\pi$ degrees,

degree = 3600''.

Multiplying these fantastic numbers together, we conclude that the rate of precession is about

$$43.1''/\text{century},$$

in perfect agreement with observation.

EXERCISES

7.1 Consider a small mass m initially at rest a huge distance R from the sun. Assuming that θ, φ remain constant, show that the relevant equations from Section 7.5 become

$$-(1 - \gamma r^{-1})^{-1}\left(\frac{dr}{d\tau}\right)^2 + (1 - \gamma r^{-1})\left(\frac{dt}{d\tau}\right)^2 = 1 \qquad \text{(I')}$$

$$(1 - \gamma r^{-1})\frac{dt}{d\tau} = \beta. \qquad \text{(IV'')}$$

(We have not made the text's assumption that $\gamma = 2GM$.) Conclude that

$$\left(\frac{dr}{dt}\right)^2 = (1 - \gamma r^{-1})^2 - \beta^{-2}(1 - \gamma r^{-1})^3.$$

Since we are assuming that initially $dr/dt = 0$, deduce that

$$f(r^{-1}) \equiv \left(\frac{dr}{dt}\right)^2 = (1 - \gamma r^{-1})^2 - (1 - \gamma r_0^{-1})^{-1}(1 - \gamma r^{-1})^3,$$

with $f'(r_0^{-1}) = \gamma(1 - \gamma r_0^{-1})$. Of course, classically $\frac{1}{2}m\,(dr/dt)^2$, the kinetic energy, equals the loss of potential energy, $GMm(1/r - 1/r_0)$, so

$$f(r^{-1}) = \left(\frac{dr}{dt}\right)^2 = 2GM\left(\frac{1}{r} - \frac{1}{r_0}\right)$$

with $f'(r_0^{-1}) = 2GM$. Assuming the theories agree asymptotically for large r_0, conclude that $\gamma = 2GM$.

8

The Gauss-Bonnet Theorem

One of the most celebrated results in mathematics, the Gauss-Bonnet Theorem, links the geometry and topology of surfaces. This chapter provides an overview without many proofs.

8.1. The Gauss-Bonnet formula. *Let R be a smooth disc in a smooth 2-dimensional Riemannian manifold M with Gauss curvature G. Let κ_g denote the geodesic curvature of the boundary. Then*

$$\int_R G + \int_{\partial R} \kappa_g = 2\pi. \tag{1}$$

For example, for a disc in the plane, $0 + 2\pi = 2\pi$. For the upper half of the unit sphere, $2\pi + 0 = 2\pi$.

Notice that this formula implies that G is intrinsic, as announced by Gauss's Theorema Egregium 3.6. The proof, like that for the Theorema Egregium, is a messy computation. It begins with a formula for G in local coordinates and changes $\int_R G$ into an integral over ∂R by Green's Theorem.

If ∂R has corners with interior angles α_i, as in Figure 8.1, then the boundary curvature term $\int_{\partial R} \kappa_g$ in (1) may be reinterpreted to include the discrete contributions $\Sigma (\pi - \alpha_i)$. Alternatively, if the

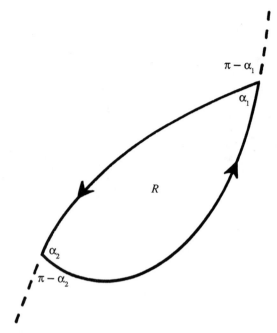

Figure 8.1. An interior angle α contributes $\pi - \alpha$ to $\int_{\partial R} \kappa_g$.

angles are treated separately,

$$\int_R G + \int_{\partial R} \kappa_g + \Sigma(\pi - \alpha_i) = 2\pi. \tag{2}$$

In particular, for a geodesic triangle $\triangle$,

$$\int_{\triangle} G + \pi = \alpha_1 + \alpha_2 + \alpha_3, \tag{3}$$

a happy variation on the familiar statement that for a planar triangle the angles sum to π. By using triangles shrinking down to a point, we may compute the Gauss curvature as

$$G = \lim \frac{\alpha_1 + \alpha_2 + \alpha_3 - \pi}{\text{area } \triangle}.$$

On the unit sphere, $\int_{\triangle} G$ becomes the area A of $\triangle$:

$$\alpha_1 + \alpha_2 + \alpha_3 = \pi + A, \tag{4}$$

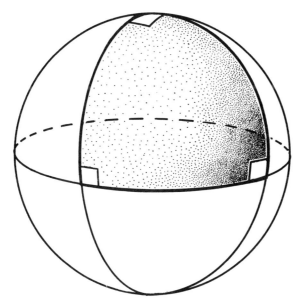

Figure 8.2.   For a spherical triangle, the sum of the angles $\alpha_1 + \alpha_2 + \alpha_3 = \pi + A$. Here $\pi/2 + \pi/2 + \pi/2 = \pi + \pi/2$.

the basic formula of spherical trigonometry. For example, for the geodesic triangle of Figure 8.2 with one vertex at the north pole, two on the equator, and three right angles,

$$\frac{\pi}{2} + \frac{\pi}{2} + \frac{\pi}{2} = \pi + \frac{\pi}{2}.$$

Gauss originally obtained formula (3) in 1827. Bonnet provided formula (1) in 1848.

8.2. The Gauss-Bonnet Theorem.   The Gauss-Bonnet Theorem is a global result about a compact, 2-dimensional smooth Riemannian manifold M. It relates a geometric quantity, the integral of the Gauss curvature, to a topological quantity, the Euler characteristic χ. For any triangulation of M, with V vertices, E edges, and F faces, χ

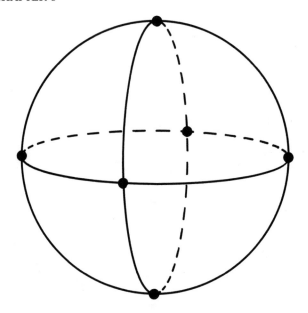

Figure 8.3.   The unit sphere has Euler characteristic $\chi = V - E + F = 6 - 12 + 8 = 2$. The Gauss-Bonnet Theorem says that $\int G = 4\pi = 2\pi\chi$.

is defined by $\chi = V - E + F$. The theorem says that

$$\int_M G = 2\pi\chi. \tag{1}$$

For example, consider the unit sphere, triangulated by the equator and two orthogonal great circles of longitude. (See Figure 8.3.) The Euler characteristic is

$$\chi = V - E + F = 6 - 12 + 8 = 2.$$

Hence

$$\int G = 4\pi = 2\pi\chi.$$

One remarkable consequence of (1) is that the Euler characteristic is independent of the choice of triangulation and hence is a topological invariant. Actually, for a surface of genus g, $\chi = 2 - 2g$.

A second remarkable consequence of (1) is that $\int G$ is independent of the metric, depending only on the topology of M.

Proof of Gauss-Bonnet Theorem.   Fix a triangulation of M. On each triangle $\triangle$, the Gauss-Bonnet formula 8.1(2) becomes

$$\int_{\triangle} G = -\int_{\partial\triangle} \kappa_g + \Sigma\alpha_i - \pi.$$

Now add up the formulas for all the triangles. The first term contributes $\int_M G$. Since each edge occurs twice in opposite directions, the various $\int_{\partial\triangle} \kappa_g$ cancel. The angles around each vertex sum to 2π, so the angle term contributes $2\pi V$. The last term contributes πF. Finally, since each face has three edges and each edge lies on two faces, $E = \frac{3}{2}F$. Therefore

$$\int_M G = 2\pi V - \pi F = 2\pi(V - \tfrac{3}{2}F + F) = 2\pi(V - E + F) = 2\pi\chi.$$

8.3.   The Gauss map of a surface in $\mathbf{R}^3$.   The *Gauss map* of a surface M in $\mathbf{R}^3$ is just the unit normal $\mathbf{n}\colon M \to S^2$. Consider such a surface as pictured in Figure 8.4, tangent to the x,y-plane at the origin p_1, with principal curvatures κ_1 along the x-axis and κ_2 along the y-axis. For the purposes of illustration, suppose $\kappa_1 < 0$ and $\kappa_2 > 0$.

We want to consider the derivative $D\mathbf{n}$, called the *Weingarten map*. If we move in the x-direction from p_1 toward a point p_2, $\mathbf{n}$ turns in the x-direction an amount proportional to $|\kappa_1|$, but positive while $\kappa_1 < 0$. Indeed, the first column of $D\mathbf{n}$ is $\begin{bmatrix} -\kappa_1 \\ 0 \end{bmatrix}$. If we move instead in the y-direction from p_1 toward a point p_3, $\mathbf{n}$ turns in the negative y-direction an amount proportional to $|\kappa_2|$. The second column of $D\mathbf{n}$ is $\begin{bmatrix} 0 \\ -\kappa_2 \end{bmatrix}$. Hence

$$D\mathbf{n} = \begin{bmatrix} -\kappa_1 & 0 \\ 0 & -\kappa_2 \end{bmatrix} = -\mathrm{II}.$$

This identity holds in any orthonormal coordinates. The Jacobian of the Gauss map equals the Gauss curvature:

$$\det D\mathbf{n} = \kappa_1\kappa_2 = G.$$

If M is a topological sphere, $\mathbf{n}$ has degree 1 (covers the sphere once, algebraically), and

$$\int_M G = \text{area image } \mathbf{n} = 4\pi = 2\pi\chi.$$

70

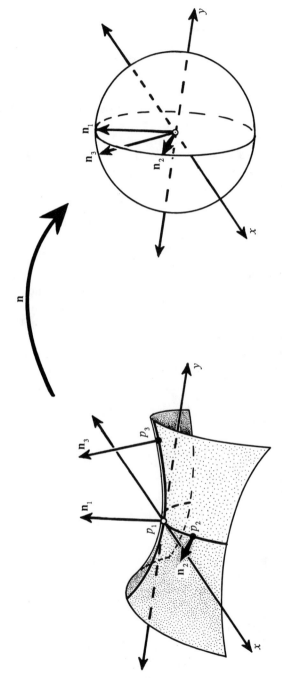

Figure 8.4. The Gauss map sends a point to the unit normal in the sphere.

We have recovered the Gauss-Bonnet Theorem for a sphere in $\mathbf{R}^3$. For general compact M in $\mathbf{R}^3$, $\mathbf{n}$ has degree $\chi/2$, and

$$\int_M G = \frac{\chi}{2} 4\pi = 2\pi\chi,$$

the Gauss-Bonnet Theorem for any closed surface in $\mathbf{R}^3$.

8.4. The Gauss map of a hypersurface. For a hypersurface M^n in $\mathbf{R}^{n+1}$, the Gauss map $\mathbf{n}: M \to S^n$. In orthonormal coordinates aligned with the principal curvature directions at a point, the Weingarten map is

$$D\mathbf{n} = \begin{bmatrix} -\kappa_1 & & 0 \\ & \ddots & \\ 0 & & -\kappa_n \end{bmatrix} = -\mathrm{II},$$

and the Jacobian of $\mathbf{n}$ is

$$(-\kappa_1) \cdots (-\kappa_n) = (-1)^n G,$$

if the Gauss curvature G is defined as $\kappa_1 \cdots \kappa_n = \det \mathrm{II}$. As for surfaces, if n is even, the degree of $\mathbf{n}$ is the Euler characteristic

$$\chi = V - E + F - \cdots$$

and

$$\int_M G = \frac{\chi}{2} \text{ area } S^n, \tag{1}$$

a generalization of Gauss-Bonnet to hypersurfaces by H. Hopf in 1925 [Ho]. (If n is odd, $\chi = 0$.)

8.5. The Gauss-Bonnet-Chern Theorem. Amazingly enough, a generalization of the Gauss-Bonnet Theorem 8.4(1) holds for any even-dimensional smooth compact Riemannian manifold M. An extrinsic proof was obtained by C. B. Allendoerfer [All] and W. Fenchel [Fen] around 1938, an intrinsic proof by S.-S. Chern [Ch] in 1944. If Nash's embedding theorem [N] says that every M can be embedded in some $\mathbf{R}^n$, why is it not called the Gauss-Bonnet-Allendoerfer-Fenchel Theorem? Because Nash's theorem was not proved until 1954.

The formulation and proof require a definition of G in local coordinates. It is

$$G = \frac{1}{2^{n/2}n! \det g_{ij}} R_{i_1 i_2 j_1 j_2} R_{i_3 i_4 j_3 j_4} \cdots R_{i_{n-1} i_n j_{n-1} j_n} \epsilon^{i_1 \cdots i_n} \epsilon^{j_1 \cdots j_n},$$

where $\epsilon^{i_1 \cdots i_n} = \pm 1$, according to whether $i_1, \ldots, i_n$ is an even or odd permutation. For example, for a 2-dimensional surface tangent to the x_1,x_2-plane at 0 in $\mathbf{R}^n$, with x_1, x_2 as local coordinates, $\det g_{ij} = 1$, and

$$G = \frac{1}{2^1 2!} (R_{1212} - R_{1221} - R_{2112} + R_{2121}) = R_{1212},$$

the Gauss curvature of the only section there is [compare to 5.2(2)].

Actually Chern used the language of differential forms and moving frames. He defined G as the Pfaffian (a square root of the determinant) of certain curvature forms. His pioneering work on fiber bundles launched the modern era in differential geometry.

8.6. Parallel transport. A vectorfield on a curve is called *parallel* if its covariant derivative along the curve vanishes [see 6.5(1)]. A vector at a point on a curve can be uniquely continued "by parallel transport" as a parallel vectorfield. In Euclidean space, a parallel vectorfield is constant—that is, the vectors are all "parallel."

In a Riemannian manifold M, a curve is a geodesic if and only if its unit tangent T is parallel. If M is a 2-dimensional surface, γ is a curve, and θ is the angle from a parallel vectorfield X to the unit tangent T, then the geodesic curvature $\kappa_g = d\theta/ds$. If γ is a closed curve, the result $X(1)$ of parallel-transporting X around the curve will be at some angle α from the starting vector $X(0)$. (See Figure 8.5.) By the Gauss-Bonnet formula 8.1(1),

$$2\pi - \int G = \int \kappa_g = \int \frac{d\theta}{ds} = 2\pi - \alpha,$$

so $\alpha = \int G$. Hence the Gaussian curvature may be interpreted as the net amount a vector turns under parallel transport around a small closed curve.

More generally, in a higher-dimensional Riemannian manifold M, R_{ijkl} may be interpreted as the amount a vector turns in the e_i,e_j-plane under parallel transport around a small closed curve in the e_k,e_l-plane.

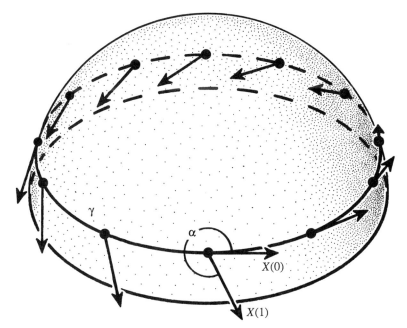

Figure 8.5.   Geodesic curvature $\kappa_g = d\theta/ds$, where θ is the angle from a parallel vectorfield X to the unit tangent **T**. By the Gauss-Bonnet formula, the angle α from the initial $X(0)$ to the final $X(1)$ equals $\int G$. For example, heading east along a circle of latitude in the northern unit hemisphere involves curving to the left (think of a small circle around the north pole). For latitude near the equator, this effect is small, and a *parallel* vectorfield ends up pointing slightly to the right, i.e., at an angle α of almost 2π to the left. Sure enough, the enclosed area also is almost 2π, the area of the whole northern hemisphere.

We have already seen the infinitesimal version of this interpretation of Riemannian curvature in formula 6.1(3):

$$X^i_{;k;l} - X^i_{;l;k} = -\sum_j R^i_{jkl} X^j.$$

The left-hand side describes the effects on X of moving in an infinitesimal parallelogram: first in the k direction, then in the l direction, then backwards along the path that went first in the l direction, then in the k direction. R^i_{jkl} gives the amount the j component of the original vector X contributes to the i component of the change.

8.7. A proof of Gauss-Bonnet in $\mathbf{R}^3$.   Ambar Sengupta has shown me a simple proof of the Gauss-Bonnet formula 8.1(1) for surfaces in $\mathbf{R}^3$. Then, of course, the Gauss-Bonnet theorem 8.2(1) follows easily as in Section 8.2.

The proof begins with a simple proof of the formula for a geodesic triangle on the unit sphere 8.1(4),

$$\alpha_1 + \alpha_2 + \alpha_3 = \pi + A, \tag{1}$$

due to Thomas Harriot (1603, see [Lo, p. 301]). The formula for a smooth disc-type region R on the sphere follows by approximation:

$$\text{area}(R) + \int_{\partial R} \kappa_g = 2\pi. \tag{2}$$

Finally, an ingenious argument deduces the formula for a smooth disc-type region on any smooth surface M in $\mathbf{R}^3$:

$$\int_R G + \int_{\partial R} \kappa_g = 2\pi. \tag{3}$$

To prove (1), consider a geodesic triangle $\triangle$ of area A and angles α_1, α_2, α_3, bounded by three great circles as in Figure 8.6. Each pair of great circles bounds two congruent lunes L_i, L'_i with angles α_i. The lunes L_i intersect in $\triangle$; the lunes L'_i intersect in a congruent triangle $\triangle'$ on the back. The lune L_i has area proportional to α_i; consideration of the extreme case $\alpha_i = \pi$ shows that area (L_i) $= 2\alpha_i$. Since $\cup L_i$ is congruent to $\cup L'_i$, each has area 2π. Hence

$$2\pi = \text{area}(\cup L_i) = \sum \text{area}(L_i) - 2A = 2(\alpha_1 + \alpha_2 + \alpha_3) - 2A.$$

Therefore $\alpha_1 + \alpha_2 + \alpha_3 = \pi + A$, as desired.

From piecing together geodesic triangles it follows that for any geodesic polygon on the sphere,

$$\text{area}(R) + \sum(\pi - \alpha_i) = 2\pi.$$

To deduce (3), consider the Gauss map $\mathbf{n}\colon M \to \mathbf{S}^2$. Since the Jacobian is the Gauss curvature G, $\mathbf{n}$ maps the region R to a region R' of area $A = \int_R G$.

Let $\gamma(t)$ $(0 \leq t \leq 1)$ be the curve bounding R. Let $X(t)$ be a parallel vectorfield on γ, so $\dot{X}(t)$ is a multiple of $\mathbf{n}$. Along the curve

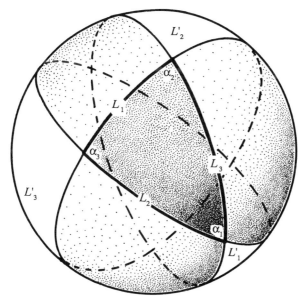

Figure 8.6.   On a unit sphere, the sum $\alpha_1 + \alpha_2 + \alpha_3$
of the angles of a geodesic triangle equals $\pi + A$, as
may be proved by viewing the triangle as the
intersection of three lunes I_{α_i}, each of area $2\alpha_i$.

$\gamma'(t) \equiv n \circ \gamma(t)$ on the sphere, the unit normal is the same, so $X(t)$,
bodily moved in $\mathbf{R}^3$ to the sphere, is still parallel.

Let α be the angle from $X(0)$ to $X(1)$. Then $\int_{\partial R} \kappa_g - 2\pi$ and
$\int_{\partial R'} \kappa_g - 2\pi$ both equal $-\alpha$ (see Figure 8.5). Therefore

$$\int_R G + \int_{\partial R} \kappa_g - 2\pi = \text{area}(R') + \int_{\partial R'} \kappa_g - 2\pi = 0$$

by (2), proving (3).

Here area(R') denotes the algebraic area of R', negative if G
is negative and **n** reverses orientation. Similarly, $\int_{\partial R'} \kappa_g - 2\pi$ must
be interpreted so that, for example, if G is negative, it switches sign
too.

9

Geodesics and Global Geometry

Our streamlined approach has avoided a deep study of geodesics or even the exponential map. This chapter discusses geodesics and some theorems that draw global conclusions from local curvature hypotheses. For example, Bonnet's Theorem 9.5 obtains a bound on the diameter of M from a bound on the sectional curvature. Cheeger and Ebin provide a beautiful reference [CE] on such topics in global Riemannian geometry.

Let M be a smooth Riemannian manifold. Recall that by the theory of differential equations, there is a unique geodesic through every point in every direction. Assume that M is (geodesically) *complete*—that is, geodesics may be continued indefinitely. (The geodesic may overlap itself, as the equator winds repeatedly around the sphere.) This condition means that M has no boundary and no missing points.

9.1. The exponential map. The *exponential map* Exp_p at a point p in M maps the tangent space T_pM into M by sending a vector $\mathbf{v}$ in T_pM to the point in M a distance $|\mathbf{v}|$ along the geodesic from p in the direction $\mathbf{v}$. (See Figure 9.1.)

For example, let M be the unit circle in the complex plane $\mathbf{C}$,

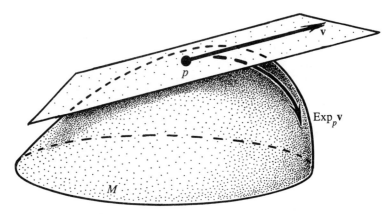

Figure 9.1. The exponential map Exp_p maps $\mathbf{v} \in T_p M$ to the point a distance $|\mathbf{v}|$ along the geodesic in the direction $\mathbf{v}$.

$p = 1$, $T_p M = \{iy\}$. Then

$$\text{Exp}_1(iy) = e^{iy}.$$

(See Figure 9.2.)

As a second example, let M be the Lie group $SO(n)$ of rotations of $\mathbf{R}^n$, represented as

$$SO(n) = \{n \times n \text{ matrices } A: AA^t = I \text{ and } \det A = 1\} \subset \mathbf{R}^{n^2}.$$

The tangent space at the identity matrix I consists of all skew-symmetric matrices,

$$T_I SO(n) = \{A: A^t = -A\},$$

because differentiating the defining relation $AA^t = I$ yields $IA^t + AI^t = 0$, that is, $A^t = -A$. (See Figure 9.3.)

The exponential map on $T_I SO(n)$ is given by the exponential matrix function familiar from linear algebra:

$$\text{Exp}_I(A) = e^A = I + A + \frac{A^2}{2!} + \cdots.$$

For any point p in a smooth Riemannian manifold M, Exp_p is a smooth diffeomorphism at 0. It provides very nice coordinates called *normal coordinates* in a neighborhood of p. Normal coordinates have the useful property that the metric $g_{ij} = I$ to first order at p. (Compare to Theorem 3.6.)

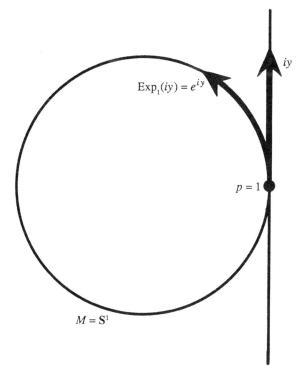

Figure 9.2. For M the circle, the tangent space is $T_1M = \{iy\}$, and the exponential map is $\text{Exp}_1(iy) = e^{iy}$.

A small open ball in normal coordinates is simple and convex: there is a unique geodesic between any two points. (*Simple* means at most one; *convex* means at least one.) Moreover, that geodesic is the shortest path in all of M between the two points.

The Hopf-Rinow Theorem says that as long as M is connected, there is a geodesic giving the shortest path between any two points. In particular, Exp_p maps T_pM onto M.

9.2. The curvature of $SO(n)$. As an example, we now compute the curvature of $SO(n)$. In Chapter 5 we defined the second fundamental tensor of a submanifold M of $\mathbf{R}^n$ by the turning rate κ of unit tangents along each slice curve. As long as we take the normal component, any curve heading in the same direction and any tangent

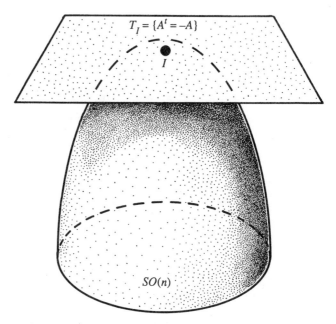

Figure 9.3. For the special orthogonal group $SO(n)$, the tangent space is $T_I SO(n) = \{A : A^t = -A\}$, and the exponential map is $\text{Exp}_I(A) = e^A$.

vectorfield starting with the same tangent vector will give the same result.

Let $\{E_i\}$ be an orthonormal basis for $T_I SO(n) = \{A = -A^t\}$. The ij component of II may be computed as the normal component of the derivative along the curve e^{sE_i} of the vectorfield $e^{sE_i}E_j$. [Since e^{sE_i} maps I to e^{sE_i}, it maps E_j to a tangent vector $(e^{sE_i})E_j$.] First we compute

$$\frac{d}{ds}(e^{sE_i}E_j) = \frac{d}{ds}(1 + sE_i + \cdots)E_j = E_i E_j.$$

The projection onto the normal space of symmetric matrices is given by

$$\tfrac{1}{2}[(E_i E_j) + (E_i E_j)^t] = \tfrac{1}{2}(E_i E_j + E_j E_i).$$

Hence

$$\text{II} = \tfrac{1}{2}(E_i E_j + E_j E_i).$$

The curvature of the E_i, E_j section is given by

$$K(E_i, E_j) = \tfrac{1}{4}[2E_i^2 \cdot 2E_j^2 - (E_iE_j + E_jE_i) \cdot (E_iE_j + E_jE_i)].$$

Since for matrices $A \cdot B = \text{trace}\,(AB^t)$ and $(AB)^t = B^tA^t$,

$$
\begin{aligned}
K(E_i, E_j) &= \tfrac{1}{4}\,\text{trace}\,(4E_iE_iE_j^tE_j^t - E_iE_jE_j^tE_i^t \\
&\quad - E_iE_jE_i^tE_j^t - E_jE_iE_j^tE_i^t - E_jE_iE_i^tE_j^t) \\
&= \tfrac{1}{4}\,\text{trace}\,(4E_iE_iE_jE_j - E_iE_iE_jE_j \\
&\quad - E_iE_jE_iE_j - E_iE_jE_iE_j - E_iE_iE_jE_j),
\end{aligned}
$$

because $E_i^t = -E_i$ and trace $(AB) = \text{trace}\,(BA)$, (although trace $(ABC) \neq \text{trace}\,(CBA)$). Hence

$$
\begin{aligned}
K(E_i, E_j) &= \tfrac{1}{4}\,\text{trace}\,(E_iE_jE_jE_i - E_iE_jE_iE_j \\
&\quad - E_jE_iE_jE_i + E_jE_iE_iE_j) \\
&= \tfrac{1}{4}\,\text{trace}\,([E_i, E_j][E_i, E_j]^t),
\end{aligned}
$$

where $[E_i, E_j]$ denotes the bracket product $E_iE_j - E_jE_i$. Therefore

$$K(E_i, E_j) - \tfrac{1}{4}\,|[E_i, E_j]|^2.$$

Indeed in any Lie group, for orthonormal vectors **v**, **w**,

$$K(\mathbf{v}, \mathbf{w}) = \tfrac{1}{4}\,|[\mathbf{v}, \mathbf{w}]|^2.$$

For example, for $SO(3)$, take

$$E_1 = \frac{1}{\sqrt{2}}\begin{bmatrix} 0 & -1 & 0 \\ 1 & 0 & 0 \\ 0 & 0 & 0 \end{bmatrix},$$

$$E_2 = \frac{1}{\sqrt{2}}\begin{bmatrix} 0 & 0 & -1 \\ 0 & 0 & 0 \\ 1 & 0 & 0 \end{bmatrix},$$

$$E_3 = \frac{1}{\sqrt{2}}\begin{bmatrix} 0 & 0 & 0 \\ 0 & 0 & -1 \\ 0 & 1 & 0 \end{bmatrix}.$$

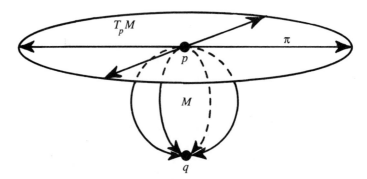

Figure 9.4. On the sphere M, Exp_p maps an open disc diffeomorphically onto $M - \{q\}$, but maps the whole boundary circle onto $\{q\}$. The singular point q for Exp_p is called a conjugate point.

Then

$$K(E_1, E_2) = \tfrac{1}{4} \left\|[E_1, E_2]\right\|^2$$

$$= \tfrac{1}{4} \left\| \begin{bmatrix} 0 & 0 & 0 \\ 0 & 0 & -\tfrac{1}{2} \\ 0 & \tfrac{1}{2} & 0 \end{bmatrix} \right\|^2 = \tfrac{1}{8}.$$

Indeed, all the sectional curvatures turn out to be $\tfrac{1}{8}$. Actually $SO(3) = \mathbf{R}P^3$ is just a round 3-sphere of radius $2\sqrt{2}$ with antipodal points identified.

For $SO(n)$, $0 \le K \le \tfrac{1}{8}$.

9.3. Conjugate points and Jacobi fields. Although Exp_p is a diffeomorphism at 0, it need not be a diffeomorphism at all points $\mathbf{v} \in T_pM$. For example, let M be the unit sphere and let p be the north pole. Then Exp_p maps the disc $\{\mathbf{v} \in T_pM : |\mathbf{v}| < \pi\}$ diffeomorphically onto $M - \{q\}$, where q is the south pole, but it maps the whole circle $\{|\mathbf{v}| = \pi\}$ onto $\{q\}$. (See Figure 9.4.)

On the other hand, for the saddle $\{z = -x^2 + y^2\}$ of Figure 9.5, Exp_0 is a global diffeomorphism.

A point $q = \text{Exp}_p\mathbf{v} \in M$ is called *conjugate* to p if Exp_p fails to be a diffeomorphism at $\mathbf{v}$—that is, if the linear map $D\,\text{Exp}_p\mathbf{v}$ is singular. This occurs when moving perpendicular to $\mathbf{v}$ at $\mathbf{v} \in T_pM$ corresponds to zero velocity at $q \in M$, or roughly when nearby

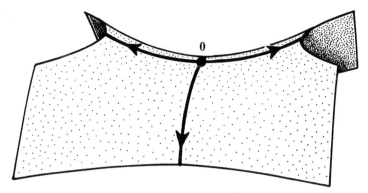

Figure 9.5. For the saddle $\{z = -x^2 + y^2\}$, Exp_0 is a global diffeomorphism.

geodesics from p focus at q. Such conjugate points q are characterized by a variation "Jacobi" vectorfield J along the geodesic, vanishing at p and q, for which the second variation of length is zero. In other words, let $\gamma_s(t)$ result from letting a finite piece of geodesic $\gamma_0(t)$ flow a distance $s|J(t)|$ in the direction $J(t)$. Let $L(s) =$ length γ_s. Then $L''(0) = 0$.

Note. It turns out that once a geodesic passes a conjugate point, it is no longer the shortest geodesic from p.

We state the following theorem as an early example of the relationship between curvature and conjugate points (see [CE, Rauch's Thm. 1.28]).

Theorem. *Let M be a smooth Riemannian manifold. If the sectional curvature K at every point for every section is bounded above by a constant K_0, then the distance from any point to a conjugate point is at least $\pi/\sqrt{K_0}$. In particular, if the sectional curvature K is nonpositive, there are no conjugate points, and Exp_p is a (local) diffeomorphism at every point. (We say that Exp_p is a submersion or a covering map.)*

9.4. Cut points and injectivity radius. A *cut point* is the last point on a geodesic from p to which the geodesic remains the shortest path from p. The cut point q could be conjugate to p, as the antipodal

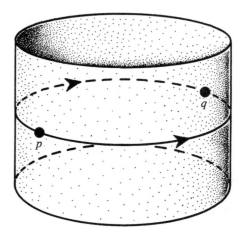

Figure 9.6. Geodesics on the cylinder originating at
p stop being shortest paths at the cut point q.

point on a sphere, where infinitesimally close geodesics focus (see
the Note on page 83). Alternatively, the cut point q could be like
the antipodal point on the cylinder of Figure 9.6, where geodesics
heading in opposite directions from p eventually meet.

Inside the locus of cut points, Exp_p is injective, a diffeomorphism. The infimum of distances from any point to a cut point is called
the *injectivity radius* of the manifold. For example, the injectivity
radius of a cylinder of radius a is πa.

Bounding the sectional curvature does not bound the injectivity
radius away from 0. A cylinder with Gauss curvature 0 can have an
arbitrarily small radius and injectivity radius. Likewise, hyperbolic
manifolds with negative curvature can have a small injectivity radius.
A common hypothesis for global theorems is *bounded geometry:*
sectional curvature bounded above and injectivity radius bounded
below.

9.5. Bonnet's Theorem.   Bonnet's Theorem draws a global conclusion from a local, curvature hypothesis:

Let M be a smooth (connected) Riemannian manifold with sec-
tional curvature bounded below by a positive constant K_0. Then the
diameter of M is at most $\pi/\sqrt{K_0}$.

The *diameter* of M is the greatest distance between any two

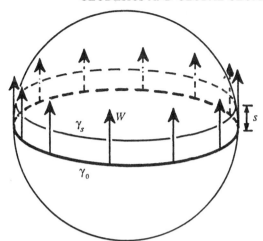

Figure 9.7. The second variation of the length of the equator $L''(0) = -\int K(T, W) = -2\pi$.

points. The unit sphere with $K = 1$ and diameter π (distance is measured on the sphere) shows that Bonnet's Theorem is sharp.

We now give a proof sketch beginning with three lemmas. The first lemma relates the second variation of the length of a geodesic to sectional curvature K.

Lemma. *Let γ be a finite piece of geodesic with unit tangent T. Let W be an orthogonal, parallel unit variational vectorfield on γ. Then the initial second variation of length is given by*

$$L''(0) = -\int K(T, W). \tag{1}$$

For example, let γ be the equator on the unit sphere of Figure 9.7. Let W be the unit upward vectorfield. Then the circle of latitude γ_s a distance s from γ has length $L(s) = 2\pi \cos s$. Hence

$$L''(0) = -2\pi = -\int K,$$

since $K = 1$.

This lemma illustrates our earlier remark (6.7) that positive curvature means that parallel geodesics converge (and hence cross-sectional distance decreases).

Note that by scaling, for the variational vectorfield aW of length a,

$$L''(0) = \int -a^2 K(T, W). \tag{2}$$

The second lemma considers variational vectorfields of variable length.

Lemma. *Let γ be an initial segment of the x-axis in $\mathbf{R}^2$ of length $L(0)$. Consider a smooth vertical variational vectorfield $f(x)\mathbf{j}$. Then the initial second variation of length is given by*

$$L''(0) = \int_0^{L(0)} f'(x)^2 \, dx. \tag{3}$$

Proof. Flowing $sf(x)\mathbf{j}$ produces a curve of length

$$L(s) = \int_0^{L(0)} [1 + s^2 f'(x)^2]^{1/2} \, dx$$

$$= \int_0^{L(0)} [1 + \tfrac{1}{2} s^2 f'(x)^2 + \cdots] \, dx.$$

Differentiation yields (3).

The third lemma states without further proof the result of combining the effects of the first two lemmas $(2, 3)$.

Lemma. *Let γ be a finite piece of geodesic with unit tangent T. Consider a variational vectorfield fW, where W is an orthogonal, parallel unit variational vectorfield on γ. Then the initial second variation of length is given by*

$$L''(0) = \int_\gamma [f'^2 - f^2 K(T, W)]. \tag{4}$$

Proof of Bonnet's Theorem. Suppose diam $M > \pi/\sqrt{K_0}$. Then there is some shortest geodesic $\gamma(t)$ of length $l > \pi/\sqrt{K_0}$. Hence $K \geq K_0 > \pi^2/l^2$.

Assume γ is parameterized by arc length t. Let W be an orthogonal, parallel unit vectorfield on γ. Take as a variational vectorfield $\left(\sin \dfrac{\pi}{l}t\right)W$, which vanishes at the endpoints of γ. By (4), the initial second variation of length is given by

$$L''(0) = \int_0^l \left(\frac{\pi}{l}\cos\frac{\pi}{l}t\right)^2 - \left(\sin^2\frac{\pi}{l}t\right)K(T, W)$$

$$< \int_0^l \frac{\pi^2}{l^2}\cos^2\frac{\pi}{l}t - \frac{\pi^2}{l^2}\sin^2\frac{\pi}{l}t = 0.$$

This contradiction of the choice of γ as a *shortest* path completes the proof.

Remark. In the proof of Bonnet's Theorem, we could have chosen any unit vector orthogonal to T for W at the starting point of γ (extending W by parallel transport). Averaging over all such choices permits us to replace the bound on K by a bound on its average, the Ricci curvature. The theorem of Myers concludes that if $\mathrm{Ric} \geq (n-1)K_0$, then $\mathrm{diam}\, M \leq \pi/\sqrt{K_0}$.

9.6. Constant curvature, the Sphere Theorem, and the Rauch Comparison Theorem. This section just mentions some famous results on a smooth, connected, complete Riemannian manifold M.

Suppose that M is simply connected, so every loop can be shrunk to a point. Suppose the sectional curvature K is constant for all sections at all points. By scaling, we may assume K is 1, 0, or -1.

If $K = 1$, M is the unit sphere. If $K = 0$, M is Euclidean space. If $K = -1$, M is hyperbolic space. Thus the metric and global geometry are completely determined for constant curvature.

The Sphere Theorem, perhaps the most famous global theorem, draws topological conclusions from hypotheses that the curvature is "pinched" between two values.

The Sphere Theorem. *Let M be simply connected with sectional curvature $\frac{1}{4} < K \leq 1$. Then M is a topological sphere (homeomorphic to the standard sphere).*

The theorem is sharp, since, for example, complex projective space $\mathbf{C}P^2$ has $\frac{1}{4} \leq K \leq 1$. It was proved by H. Rauch for $\frac{3}{4} \leq K \leq 1$ in 1951, and generalized to $\frac{1}{4} < K \leq 1$ by M. Berger and W. Klingenberg in 1960.

The Rauch Comparison Theorem.   One of the main ingredients in a proof, and one of the most useful tools in Riemannian geometry, is the Rauch Comparison Theorem. It says, for example, the following:

Let M_1, M_2 be complete smooth Riemannian manifolds with sectional curvatures $K_1 \geq K_0 \geq K_2$ for some constant K_0. For $p_1 \in M_1$, $p_2 \in M_2$, identify $T = T_{p_1}M_1 = T_{p_2}M_2$ via a linear isometry. Let B be an open ball about 0 in T on which Exp_{p_1} and Exp_{p_2} are diffeomorphisms into M_1 and M_2. Let γ be a curve in B, and let γ_1, γ_2 be its images in M_1, M_2. Then length $(\gamma_1) \leq$ length (γ_2).

In applications, either M_1 or M_2 is usually taken to be a sphere, Euclidean space, or hyperbolic space, all of which have well-known trigonometries. Thus one obtains distance estimates on the other manifold from curvature bounds.

10

General Norms

In nature, the energy of a path or surface often depends on direction as well as length or area. The surface energy of a crystal, for example, depends radically on direction. Indeed, some directions are so much cheaper that most crystals use only a few cheap directions. (See Figure 10.1.) This chapter applies more general costs or norms Φ to curves and presents an appropriate generalization of curvature.

10.1 Norms. A norm Φ on $\mathbf{R}^n$ is a nonnegative, convex homogeneous function on $\mathbf{R}^n$. We call Φ C^k if its restriction to $\mathbf{R}^n - \{0\}$ is C^k (or, equivalently, if its restriction to the unit sphere $\mathbf{S}^{n-1}$ is C^k). The convexity of Φ is equivalent to the convexity of its unit ball

$$\{x: \Phi(x) \leq 1\}.$$

For any curve C, parametrized by a differentiable map $\gamma: [0, 1] \to \mathbf{R}^n$, define

$$\Phi(C) = \int_C \Phi(\mathbf{T}) \, ds = \int_{[0,1]} \Phi(\dot{\gamma}) \, dt.$$

If C is a straight line segment, then

$$\Phi(C) = \Phi(\mathbf{T}) \text{ length } C.$$

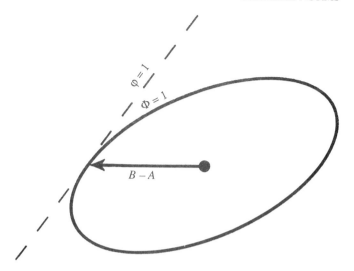

Figure 10.2. Since the unit ball of Φ is strictly convex, there is a linear function or 1-form φ such that $\varphi(v) \le \Phi(v)$, with equality only if $v = B - A$.

10.2. Proposition. *Among all differentiable curves C from A to B, the straight line L minimizes $\Phi(C)$ uniquely if Φ is strictly convex.*

Proof. Since the unit ball of Φ is convex, there is a constant-coefficient differential form φ such that

$$\varphi(v) \le \Phi(v),$$

with equality when $v = B - A$. (See Figure 10.2.) If Φ is strictly

---

Figure 10.1. Crystal shapes typically have finitely many flat facets corresponding to surface orientation of low energy. (The first two photographs are from Steve Smale's Beautiful Crystals Calendar; current version available for $12 from 69 Highgate Road, Kensington, CA 94707. The third photograph is from E. Brieskorn. All three appeared in *Mathematics and Optimal Form* by S. Hildebrandt and A. Tromba [HT, p. 181].)

convex, equality holds only if v is a multiple of $B - A$. Let C' be any differentiable curve from A to B. Then

$$\Phi(C') = \int_{C'} \Phi(\mathbf{T})\, ds \geq \int_{C'} \varphi\, ds$$

$$= \int_{C} \varphi\, ds = \Phi(C)$$

by Stokes's Theorem, so C is Φ-minimizing. If Φ is strictly convex, the inequality is strict unless C' is also a straight line from A to B, so C is uniquely minimizing.

10.3. Proposition. *A nonnegative homogeneous C^2 function Φ on $\mathbf{R}^n$ is convex (respectively, uniformly convex) if and only if the restrictions $\Phi(\theta)$ of Φ to circles about the origin satisfy*

$$\Phi''(\theta) + \Phi(\theta) \leq 0 \quad (<0).$$

Proof. Since convexity in every plane through 0 is equivalent to convexity, we may assume $n = 2$. The curvature κ of any graph $r = f(\theta)$ in polar conditions is given by

$$\kappa = \frac{f^2 - ff'' + 2f'^2}{(f^2 + f'^2)^{3/2}}.$$

Therefore the curvature of the boundary of the unit ball $r = 1/\Phi(\theta)$ is given by

$$\kappa = \left(\frac{\Phi}{\sqrt{\Phi^2 + \Phi'^2}} \right)^3 (\Phi + \Phi'').$$

The proposition follows.

10.4. Generalized curvature. *Let C be a C^2 curve with arc length parametrization $f: [0, 1] \to \mathbf{R}^n$ and curvature vector κ. Let Φ be a C^2 norm. Consider variations δf supported in $(0, 1)$. Then the first variation satisfies*

$$\delta\Phi(f) = -\int_{[0,1]} D^2\Phi(\kappa) \cdot \delta f\, ds,$$

where $D^2\Phi$ represents the second derivative matrix evaluated at the unit tangent vector. In particular, for the case of length ($\Phi(x) = L(x) \equiv |x|$),

$$\delta L(f) = -\int_{[0,1]} \kappa \cdot \delta f \, ds.$$

In general, we call $D^2\Phi(\kappa)$ the *generalized Φ-curvature vector.*

Proof. Since $\Phi(f) = \int \Phi(f'(u)) \, du$ for any parameterization $f(u)$,

$$\delta\Phi(f) = \int D\Phi(f') \cdot \delta f'(u) \, du$$

$$= -\int D^2\Phi(f')(f'') \cdot \delta f(u) \, du$$

by integration by parts. Since for the initial arc length parametrization, f' is the unit tangent vector and f'' is the curvature vector κ, initially

$$\delta\Phi(f) = -\int D^2\Phi(\kappa) \cdot \delta f(s) \, ds.$$

10.5. The isoperimetric problem.   One famous isoperimetric theorem says that among all closed curves C in $\mathbf{R}^n$ of fixed length, the circle bounds the most area—that is, the oriented area-minimizing surface S of greatest area (see, for example, [F, 4.5.14]). In other words, an area-minimizing surface S with given boundary C satisfies

$$\text{area } S \leq \frac{1}{4\pi}(\text{length } C)^2.$$

Given a convex norm Φ or $\mathbf{R}^n$, we seek a closed curve C_0 of fixed cost $\Phi(C_0)$ which bounds the most area, so any area-minimizing surface S with given boundary C satisfies

$$\text{area } S \leq \alpha[\Phi(C)]^2,$$

with equality for $C = C_0$.

Almgren's methods [Alm, esp. Section 9] using geometric measure theory show that such an *optimal isoperimetric curve* exists.

In the plane such curves have a nice characterization. Let Ψ be a 90° rotation of Φ, so

$$\Phi(C) = \int_C \Psi(\mathbf{n}),$$

where $\mathbf{n}$ is the unit normal obtained by rotating the unit tangent $\mathbf{T}$ 90° counterclockwise. The dual norm Ψ^* is defined by

$$\Psi^*(w) = \sup\{v \cdot w \colon \Psi(v \le 1\},$$

so $|v \cdot w| \le \Psi(v)\,\Psi(w)$. The optimal isoperimetric curve is simply the boundary of the unit Φ^*-ball or "Wulff shape" (see [Wu] or [T1]). Here we sketch a short new proof, based on Schwartz symmetrization, as recently used by Brothers [Br] and Gromov [BG, Section 6.6.9, p. 215] and earlier by Knothe [K]. The same result and proof hold for optimal isoperimetric hypersurfaces in all dimensions.

10.6. Theorem.   *Let Ψ be a norm on $\mathbf{R}^2$. Among all curves enclosing the same area, the boundary of the unit Ψ^*-ball B (Wulff shape) minimizes $\int_{\partial B} \Psi(\mathbf{n})$.*

Proof sketch.   Consider any planar curve enclosing a region B' of the same area as B. Let f be an area-preserving map from B' to B carrying vertical lines linearly to vertical lines. Then $\det Df = 1$ and Df is triangular:

$$Df = \begin{bmatrix} a & 0 \\ * & b \end{bmatrix}.$$

Since $\det Df = ab = 1$, $\operatorname{div} f = a + b \ge 2$. Hence

$$\Phi(\partial B') = \int_{\partial B'} \Psi(\mathbf{n}) \ge \int_{\partial B'} \Psi(\mathbf{n})\Psi^*(f)$$

$$\ge \int_{\partial B'} f \cdot \mathbf{n} = \int_{B'} \operatorname{div} f \ge 2 \text{ area } B' = 2 \text{ area } B,$$

with equality if $B' = B$.

Remark.   Careful attention to the inequalities in the proof recovers the result of J. Taylor [T2] that the Wulff shape is the unique minimizer among measurable sets [BrM].

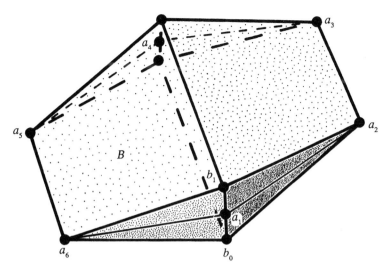

Figure 10.3. The unit Φ-ball B. Any slice A by a plane P through the origin has a vertex a_1 that is not a vertex of B.

In general dimensions, optimal isoperimetric curves are not well understood. Obvious candidates are planar Wulff shapes. The following new theorem says, however, that optimal isoperimetric curves are not generally planar.

10.7. Theorem.   *For some convex norms* Φ *in* $\mathbf{R}^3$, *an optimal isoperimetric curve is nonplanar.*

Proof.   Define Φ by taking the unit Φ-ball to be the centrally symmetric polyhedron B of Figure 10.3. In any plane P, which we may translate to pass through the origin, the Wulff shape S with boundary C maximizes (area S)/$\Phi(C)^2$. We will show that this ratio is larger for some nonplanar curve C.

The slice A of the unit Φ-ball B by the plane P must be polygonal. At least one vertex a_1 is not a vertex of B. The vertex a_1 must lie on an edge of B, with vertices b_0, b_1. The Wulff shape S is the polygon formed dual to A, rotated clockwise 90°. (See Figure

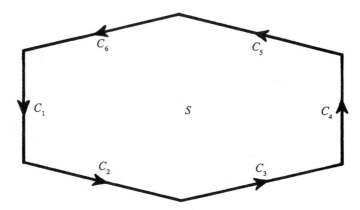

Figure 10.4.   In the plane P, the Wulff crystal S is the polygon dual to A, rotated 90° clockwise.

10.4.) If C_1 denotes the edge dual to a_1 (rotated 90°), then C_1 points in the a_1 direction. Its distance from the origin is $1/|a_1|$.

Let C' be the polygon in $\mathbf{R}^3$ obtained from C by replacing C_1 by two segments in the directions b_0, b_1, in the order that keeps the projection PC' of C' onto P out of the interior of C. (See Figure 10.5.) Then $\Phi(C') = \Phi(C)$. Let S' be an area-minimizing surface bounded by C'. Then

$$\text{area } S' > \text{area } PS' > \text{area } S.$$

Consequently,

$$\frac{\text{area } S'}{\Phi(C')^2} > \frac{\text{area } S}{\Phi(C)^2}.$$

Remark.   By approximation one obtains examples that are also smooth and elliptic.

For length, optimal isoperimetric curves are circles of constant curvature. For general Φ, the generalized curvature at least satisfies an inequality.

10.8. Lemma.   *For a C^2 optimal isoperimetric curve C_0, the generalized Φ-curvature vector satisfies*

$$|D^2\Phi(\kappa)| \le \frac{\Phi(C_0)}{2 \text{ area } S_0}. \tag{1}$$

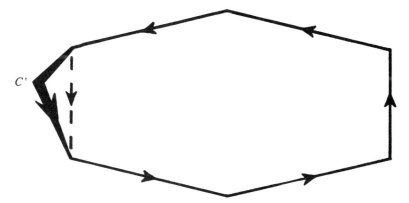

Figure 10.5.   Obtain C' from C by replacing C_1 with segments in the directions b_0, b_1. Then $\Phi(C') = \Phi(C)$, but area $S' >$ area S.

Remarks.   For the case where Φ is length and C_0 is the unit circle, (1) says $|\kappa| \leq 1$. The smoothness hypothesis on C_0 is unnecessary; still the conclusion implies that C_0 is $C^{1,1}$. If C_0 bounds a unique smooth area-minimizing surface S_0 with $\mathbf{n}$ the inward normal to C_0 along S_0, $D^2\Phi(\kappa)$ actually must be a constant multiple of $\mathbf{n}$. In particular, a planar optimal isoperimetric curve has constant generalized curvature:

$$|D^2\Phi(\kappa)| = K.$$

Proof.   Let $f : [0, a] \to \mathbf{R}^n$ be a local arc length parameterization of C_0. Consider compactly supported variations δf. Then

$$0 \geq \delta(\text{area } S - \alpha\Phi(C)^2)$$

$$\geq -\int |\delta f| \, ds + 2\alpha\Phi(C_0) \int D^2\Phi(\kappa) \cdot \delta f \, ds$$

by 10.4. Therefore

$$|D^2\Phi(\kappa)| \leq \frac{1}{2\alpha\Phi(C_0)} = \frac{\Phi(C_0)}{2 \text{ area } S_0}.$$

A norm Φ is called *crystalline* if the unit Φ-ball is polytope.

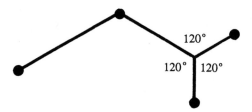

Figure 10.6.   Length-minimizing networks meet in threes at equal angles of 120°.

10.9. Conjecture.   *If Φ is crystalline, then an optimal isoperimetric curve is a polygon.*

10.10. Φ-minimizing networks.   A *network N* is a finite collection of line segments. Given a norm Φ and a finite set of boundary points in $\mathbf{R}^n$, we seek a Φ-minimizing network connecting the points. For the case where Φ is length (the generalized "Steiner" or "Fermat" problem [Fer, 1638, p. 153], [S1, 1835], [S2, 1837], [JK, 1934]), such networks meet only in threes at equal 120° angles (or in twos at boundary points at angles of at least 120°), as shown in Figure 10.6. Soap film strips behave similarly in their quest to minimize area, as shown in Figure 10.7 (see also [CR, pp. 354–361, 392]).

Recently there have appeared results on general norms Φ. (See the surveys [M1–3] and [A, GM].)

10.11. Theorem (A. Levy, Williams undergraduate '88; [Le; A3]).
Let Φ be a differentiable, uniformly convex norm on $\mathbf{R}^2$. Then Φ-minimizing networks meet only in threes.

The proof shows that a junction of four or more segments is unstable.

10.12. Theorem [LM, Theorem 4.4].   *Let Φ be a differentiable norm on $\mathbf{R}^n$. In Φ-minimizing networks, $n + 1$ segments can meet at a point, but never $n + 2$.*

It turns out that all such junctions locally can be "calibrated" and classified.

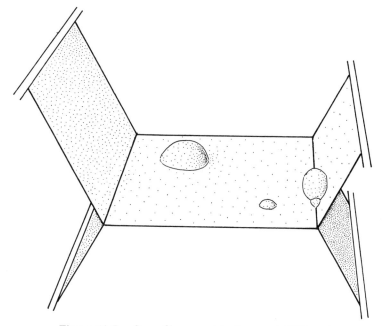

Figure 10.7. Soap films meet in threes at 120° angles in an attempt to minimize area.

The next theorem, in comparison with Theorem 10.11, exhibits a surprising sensitivity to smoothness class.

10.13. Theorem (SMALL Geometry Group, Williams, summer 1988, [A3, A4]). *Consider piecewise differentiable, uniformly convex norms* Φ *on* **R**². *Then* Φ*-minimizing networks sometimes meet in fours, although never in fives.*

The proof shows that an X can be Φ-minimizing by symmetry arguments and calculus. The proof is much easier for the "rectilinear norm" or "Manhattan metric" $Φ_M$, which is not *uniformly* convex (see [Ha]).

E. Cockayne [Coc] earlier studied planar norms, but did not discuss dependence on differentiability.

10.14. Theorem (M. Conger, Williams '89; [Con]). *Consider piecewise differentiable, uniformly convex norms* Φ *on* **R**³. *Then* Φ*-minimizing networks sometimes meet in sixes.*

The proof shows that six segments meeting along orthogonal axes in $\mathbf{R}^3$ are Φ-minimizing for some Φ. The large number of possible competitors requires cleverness as well as persistence in the proof.

Conger conjectured that the sharp bound for the number of segments meeting in a Φ-minimizing network in $\mathbf{R}^n$ is $2n$.

For non–uniformly convex norms, if the unit Φ-ball is a cube in $\mathbf{R}^n$, the network consisting of the 2^n segments from the center to the vertices is Φ-minimizing, with an easy proof. This probably exhibits the upper bound (see [FLM, Intro. and 2.1]).

More recent work by M. Alfaro (Williams '90) and others [A1, A2] has studied directed networks of one-way streets.

Selected Formulas

Curvature vector:

$$\boldsymbol{\kappa} = d\,\mathbf{T}/ds \qquad\qquad 2.0(1)$$

Metric of surface $\mathbf{x}(u^i)$:

$$g_{ij} \equiv \mathbf{x}_i \cdot \mathbf{x}_j \qquad\qquad 3.4(2)$$

$(u^1, u^2, \ldots$ give parameters on surface; $\mathbf{x}_i \equiv \partial\mathbf{x}/\partial u^i)$

Inverse matrix g^{ij}

Arc length of curve $u(t)$:

$$\int \sqrt{\sum g_{ij}\dot{u}^i\dot{u}^j}\; dt \qquad\qquad 3.4(1)$$

$(\dot{u}^i \equiv du^i/dt)$

2-dimensional surface $\mathbf{x}(u^1, u^2)$ in $\mathbf{R}^3$:

Mean curvature $H = \text{trace II} = \kappa_1 + \kappa_2$

$$H = \frac{\mathbf{x}_2^2\mathbf{x}_{11} - 2(\mathbf{x}_1 \cdot \mathbf{x}_2)\mathbf{x}_{12} + \mathbf{x}_1^2\mathbf{x}_{22}}{\mathbf{x}_1^2\mathbf{x}_2^2 - (\mathbf{x}_1 \cdot \mathbf{x}_2)^2} \cdot \mathbf{n} \qquad\qquad 3.5(1)$$

Gauss curvature $G = \det \text{II} = \kappa_1\kappa_2$

$$G = \frac{(\mathbf{x}_{11} \cdot \mathbf{n})(\mathbf{x}_{22} \cdot \mathbf{n}) - (\mathbf{x}_{12} \cdot \mathbf{n})^2}{\mathbf{x}_1^2\mathbf{x}_2^2 - (\mathbf{x}_1 \cdot \mathbf{x}_2)^2} \qquad\qquad 3.5(2)$$

For the graph of a function $f: \mathbf{R}^2 \to \mathbf{R}$,

$$H = \frac{(1 + f_y^2)f_{xx} - 2f_xf_yf_{xy} + (1 + f_x^2)f_{yy}}{(1 + f_x^2 + f_y^2)^{3/2}} \qquad\qquad 3.5(3)$$

$$G = \frac{f_{xx}f_{yy} - f_{xy}^2}{(1 + f_x^2 + f_y^2)^2} \qquad\qquad 3.5(4)$$

2-dimensional surface $x(u^1, u^2)$ in $\mathbf{R}^n$:

Mean curvature vector $\mathbf{H}$ = trace II

$$\mathbf{H} = P\frac{\mathbf{x}_2^2\mathbf{x}_{11} - 2(\mathbf{x}_1 \cdot \mathbf{x}_2)\mathbf{x}_{12} + \mathbf{x}_1^2\mathbf{x}_{22}}{\mathbf{x}_1^2\mathbf{x}_2^2 - (\mathbf{x}_1 \cdot \mathbf{x}_2)^2} \qquad 4.2(1)$$

Gauss curvature G = det II

$$G = \frac{(P\mathbf{x}_{11}) \cdot (P\mathbf{x}_{22}) - (P\mathbf{x}_{12})^2}{\mathbf{x}_1^2\mathbf{x}_2^2 - (\mathbf{x}_1 \cdot \mathbf{x}_2)^2}, \qquad 4.2(2)$$

where P denotes projection onto $T_pS^{\perp}$.

For the graph of a function $f: \mathbf{R}^{n-1} \to \mathbf{R}$,

$$H = \text{div}\frac{\nabla f}{\sqrt{1 + |\nabla f|^2}} = \frac{(1 + |\nabla f|^2)\,\Delta f - \Sigma f_i f_j f_{ij}}{(1 + |\nabla f|^2)^{3/2}},$$

where $f_i \equiv \partial f/\partial x_i$, $f_{ij} \equiv \partial^2 f/\partial x_i\,\partial x_j$, $\nabla f \equiv (f_1, \ldots, f_{n-1})$, div $(p, q, \ldots) \equiv p_1 + q_2 + \cdots$, and $\Delta f \equiv$ div $\nabla f = f_{11} + f_{22} + \cdots$ (Exercise 5.3).

For the level set $\{f = c\}$ of a function $f: \mathbf{R}^n \to \mathbf{R}$,

$$H = -\,\text{div}\frac{\nabla f}{|\nabla f|} \qquad 5.0(1)$$

Christoffel symbols:

$$\Gamma^i_{jk} = \tfrac{1}{2}\sum_l g^{il}(g_{lj,k} + g_{lk,j} - g_{jk,l}) \qquad 6.0(2)$$

Riemannian curvature tensor:

$$R^i_{jkl} = a_{ik} \cdot a_{jl} - a_{jk} \cdot a_{il} \qquad 5.2(3)$$

$$R^i_{jkl} = -\Gamma^i_{jk,l} + \Gamma^i_{jl,k} + \sum_h (-\Gamma^h_{jk}\Gamma^i_{hl} + \Gamma^h_{jl}\Gamma^i_{hk}) \qquad 6.0(4)$$

(a_{jk} are components of second fundamental form II in orthonormal coordinates.)

Ricci curvature:

$$R_{jl} = \sum_i R^i_{jil} \qquad 6.0(7)$$

Scalar curvature:

$$R = \sum g^{jl}R_{jl} \qquad 6.0(8)$$

(If S is 2-dimensional, $G = R/2$.)

Sectional curvature for v, w orthonormal:

$$K(v \wedge w) = \mathrm{II}(v, v) \cdot \mathrm{II}(w, w) - \mathrm{II}(v, w)^2 \qquad 5.2(1)$$

$$= \sum g_{ih} R^h_{jkl} v^i w^j v^k w^l. \qquad 6.0(9)$$

Covariant derivative of a vectorfield X^i:

$$X^i_{;j} = X^i_{,j} + \sum_k \Gamma^i_{jk} X^k \qquad 6.0(1)$$

Geodesics $u(t)$, t arc length:

$$0 = \ddot{u}^i + \sum_{j,k} \Gamma^i_{jk} \dot{u}^j \dot{u}^k \qquad 6.5(2)$$

Gradient:

$$\nabla f^i = g^{ij} f_{,j}$$

Laplacian:

$$\Delta f = g^{ij}(f_{,ij} - \Gamma^k_{ij} f_{,k}) = \frac{1}{\sqrt{\det g}} \frac{\partial}{\partial u_i} (\sqrt{\det g}\, g^{ij} f_{,j})$$

Bibliography

[A]   C. Adams, D. Bergstrand, and F. Morgan. The Williams SMALL undergraduate research project. *UME Trends*, January, 1991.

[A1]   Manuel Alfaro. Existence of shortest directed networks in $\mathbf{R}^2$. *Pacific J. Math.* (in press).

[A2]   Manuel Alfaro, Tessa Campbell, Josh Sher, and Andres Soto. Length-minimizing directed networks can meet in fours. SMALL Undergraduate Research Project, Williams College, 1989.

[A3]   Manuel Alfaro, Mark Conger, Kenneth Hodges, Adam Levy, Rajiv Kochar, Lisa Kuklinski, Zia Mahmood, and Karen von Haam. Segments can meet in fours in energy-minimizing networks. *J. Und. Math.* 22 (1990), 9–20.

[A4]   Manuel Alfaro, Mark Conger, Kenneth Hodges, Adam Levy, Rajiv Kochar, Lisa Kuklinski, Zia Mahmood, and Karen von Haam. The structure of singularities in Φ-minimizing networks in $\mathbf{R}^2$. *Pacific J. Math.* 149 (1991), 201–210.

[All]   C. B. Allendoerfer. The Euler number of a Riemannian manifold. *Amer. J. Math.* 62 (1940), 243–248.

[Alm]   F. Almgren. Optimal isoperimetric inequalities. *Indiana U. Math. J.* 35 (1986), 451–547.

[BG]   Marcel Berger and Bernard Gostiaux. *Differential Geometry: Manifolds, Curves, and Surfaces*. New York: Springer-Verlag, 1988.

[Br]   John E. Brothers. The isoperimetric theorem. Research report, Australian Natl. Univ. Cent. Math. Anal., January, 1987.

[BrM]   John E. Brothers and Frank Morgan. The isoperimetric theorem for general integrands, preprint (1992).

[CE]   Jeff Cheeger and David G. Ebin. *Comparison Theorems in Riemannian Geometry*. Amsterdam: North-Holland, 1975.

[Ch]   S.-S. Chern. A simple intrinsic proof of the Gauss-Bonnet formula for closed Riemannian manifolds. *Ann. of Math.* 45 (1944), 747–752.

[Coc]   E. J. Cockayne. On the Steiner problem. *Canadian Math. Bull.* 10 (1967), 431–450.

[Con]   Mark A. Conger. Energy-minimizing networks in R^n. Honors thesis, Williams College, 1989, expanded 1989.

[Cos]   C. Costa. Imersoes minimas completas em R^3 de genero um e curvatura total finita. Doctoral thesis, IMPA, Rio de Janeiro, Brazil, 1982. (See also Example of a complete minimal immersion in R^3 of genus one and three embedded ends, *Bull. Math. Soc. Bras. Mat.* 15 (1984), 47–54.)

[CR]   R. Courant and H. Robbins. *What Is Mathematics?* New York: Oxford Univ. Press, 1941.

[F]   Herbert Federer. *Geometric Measure Theory.* New York: Springer, 1969.

[Fen]   W. Fenchel. On total curvature of Riemannian manifolds: I. *J. London Math. Soc* 15 (1940), 15–22.

[Fer]   P. Fermat. *Oeuvres de Fermat.* Paris: Gauthier-Villars, 1891.

[FLM]   Z. Füredi, J. C. Lagarias, and F. Morgan. Singularities of minimal surfaces and networks and related extremal problems in Minkowski space. *Proc. DIMACS Series in Discrete Math. and Comp. Sci.* 6 (1991), 95–106.

[GM]   Thomas Garrity and Frank Morgan. SMALL Undergraduate Mathematics Research Project, Williams College, preprint (1991).

[Ha]   M. Hanan. On Steiner's problem with rectilinear distance. *J. SIAM Appl. Math.* 14 (1966), 255–265.

[He]   Sigurdur Helgason. *Differential Geometry, Lie Groups, and Symmetric Spaces.* Boston: Academic Press, 1978.

[Hi]   Noel J. Hicks. *Notes on Differential Geometry.* Princeton: Van Nostrand, 1965 (out of print).

[HT]   Stefan Hildebrandt and Anthony Tromba. *Mathematics and Optimal Form.* New York: Scientific American Books, Inc., 1985.

[Ho]   David Hoffman. The computer-aided discovery of new embedded minimal surfaces. *Math. Intelligencer* 9 (1987), 8–21.

[HoM]   D. Hoffman and W. H. Meeks III. A complete embedded minimal surface in R^3 with genus one and three ends. *J. Differential Geom.* 21 (1985), 109–127.

[JK]   V. Jarnik and O. Kossler. O minimalnich grafesh obsahujicich n danych bodu. *Casopis Pest. Mat. Fys.* 63 (1934), 223–235.

[Je]   G. B. Jeffery. *Relativity for Physics Students*. New York: Dutton, 1924.

[K]   Herbert Knothe. Contributions to the theory of convex bodies. *Mich. Math. J.* 4 (1957), 39–52.

[L]   Detlef Laugwitz. *Differential and Riemannian Geometry*, Boston: Academic Press, 1965.

[LM]   Gary Lawlor and Frank Morgan. Paired calibrations applied to soap films, immiscible fluids, and surfaces or networks minimizing other norms. Preprint (1990).

[Le]   Adam Levy. Energy-minimizing networks meet only in threes. *J. Und. Math.* 22 (1990), 53–59.

[Lo]   J. A. Lohne. Essays on Thomas Harriot. *Arch. Hist. Ex. Sci.* 20 (1979), 189–312.

[M]   Frank Morgan. *Geometric Measure Theory: A Beginner's Guide*. Boston: Academic Press, 1988.

[M1]   Frank Morgan. Compound soap bubbles, shortest networks, and minimal surfaces. AMS video (1992).

[M2]   Frank Morgan. Minimal surfaces, crystals, and norms on $\mathbf{R}^n$. In *Computational Geometry* (Proc. 7th Annual Symposium on Computational Geometry). ACM Press, 1991.

[M3]   Frank Morgan. Minimal surfaces, crystals, shortest networks, and undergraduate research. *Math. Intelligencer* (in press).

[M4]   Frank Morgan. Calculus, planets, and general relativity. *SIAM Review* 34 (1992).

[N]   J. Nash. The embedding problem for Riemannian manifolds. *Ann. of Math* (2) 63 (1956), 20–63.

[Sc]   Richard Schoen. Conformal deformation of a Riemannian metric to constant scalar curvature. *J. Diff. Geom.* 20 (1984), 479–495.

[Sp]   Barry Spain. *Tensor Calculus*. New York: Wiley, 1953.

[S]   Michael Spivak. *A Comprehensive Introduction to Differential Geometry*, Vol. I–V. Houston: Publish or Perish, 1979.

[S1]   Jacob Steiner. Aufgaben und Lehrsatze. *Crelle's Journal* 13 (1835), 361–365.

[S2]   Jacob Steiner. *Uber den Punct der kleinsten Entfernung* (1837). Berlin: Gesammelte Werke, 1882.

[St]   J. J. Stoker. *Differential Geometry*. New York: Wiley-Interscience, 1969.

[T1]   Jean E. Taylor. Crystalline variational problems. *Bull. Amer. Math. Soc.* 84 (1978), 568–588.

[T2]   Jean E. Taylor. Unique structure of solutions to a class of nonelliptic variational problems. *Proc. Symp. Pure Math.* 27 (1975), 419–427.

[W]   Steven Weinberg. *Gravitation and Cosmology*. New York: Wiley, 1972.

[Wu]   G. Wulff. Zur Frage der Geschwindigneit des Wachsthums und der Auflösung der Krystullflachen. *Zeitschrift für Krystallographi und Mineralogie* 34 (1901), 449–530.

Solutions to Selected Exercises

3.1. a. $1/a$, $1/a$, $1/a^2$, $2/a$ (or $-1/a$, $-1/a$, $1/a^2$, $-2/a$)

b. $2a$, $2b$, $4ab$, $2(a + b)$

c. $\text{II} = D^2 f(0, 0) = \begin{bmatrix} 132 & -24 \\ -24 & 118 \end{bmatrix}$,

$H = 132 + 118 = 250$,

$G = 132 \cdot 118 - 24^2 = 15{,}000$,

$\kappa = \dfrac{H \pm \sqrt{H^2 - 4G}}{2} = 100{,}150.$

d. Note that the x,y-plane is not the tangent plane at $\mathbf{0}$, so use 3.5(3,4).

$H = 4\sqrt{2}$, $G = 6$. Hence, $\kappa = 3\sqrt{2}$, $\sqrt{2}$.

e. Switch variables and use 3.5(3,4). $H = 0$,

$G = -1/(1 + y^2 \sec^2 z)^2 = -1/(1 + x^2 + y^2)^2$,

$\kappa = \pm\, 1/(1 + y^2 \sec^2 z) = \pm\, 1/(1 + x^2 + y^2).$

f. Use 3.5(3,4) and implicit differentiation.

$H = \dfrac{5 \cdot 3^4 x^2 + 10 \cdot 2^4 y^2 + 13 z^2}{(81x^2 + 16y^2 + z^2)^{3/2}},$

$G = \left(\dfrac{36}{81x^2 + 16y^2 + z^2} \right)^2,$

$\kappa = \dfrac{H \pm \sqrt{H^2 - 4G}}{2}.$

3.2. $2\pi a \sin c$. (It is a circle of radius $a \sin c$.)

3.3. $\mathbf{x} = (x, y, f(x, y))$, $\mathbf{n} = \dfrac{(-f_x, -f_y, 1)}{\sqrt{1 + f_x^2 + f_y^2}}$,

$$H = \frac{(1 + f_y^2)f_{xx} - 2f_x f_y f_{xy} + (1 + f_x^2)f_{yy}}{(1 + f_x^2)(1 + f_y^2) - f_x^2 f_y^2} \frac{1}{\sqrt{1 + f_x^2 + f_y^2}},$$

$$G = \frac{f_{xx}f_{yy} - f_{xy}^2}{1 + f_x^2 + f_y^2} \frac{1}{1 + f_x^2 + f_y^2}.$$

3.4. $\mathbf{x}_z = (f' \cos \theta, f' \sin \theta, 1)$, $\mathbf{x}_\theta = (-f \sin \theta, f \cos \theta, 0)$,

$\mathbf{n} = -(\cos \theta, \sin \theta, -f')(1 + f'^2)^{-1/2}$ (inward),

$\mathbf{x}_{zz} = (f'' \cos \theta, f'' \sin \theta, 0)$, $\mathbf{x}_{z\theta} = (-f' \sin \theta, f' \cos \theta, 0)$,

$\mathbf{x}_{\theta\theta} = (-f \cos \theta, -f \sin \theta, 0)$,

$$H = \frac{-f^2 f'' + f(1 + f'^2)}{f^2(1 + f'^2)} \frac{1}{\sqrt{1 + f'^2}} = \kappa + \frac{1}{f\sqrt{1 + f'^2}}.$$

3.5. $A = \displaystyle\int_{\varphi=0}^{r/a} 2\pi a \sin \varphi \, a \, d\varphi = 2\pi a^2 \left(1 - \cos \dfrac{r}{a}\right) = 2\pi r^2 - \dfrac{1}{a^2} \dfrac{\pi}{12} r^4 + \cdots$

4.1. a. $\mathbf{x} = (x, y, x^2 + 2y^2, 66x^2 - 24xy + 59y^2)$,

$\mathbf{x}_1 = (1, 0, 0, 0)$, $\mathbf{x}_2 = (0, 1, 0, 0)$, $\mathbf{x}_{11} = (0, 0, 2, 132)$,

$\mathbf{x}_{12} = (0, 0, 0, -24)$, $\mathbf{x}_{22} = (0, 0, 4, 118)$. $P\mathbf{x}_{ij} = \mathbf{x}_{ij}$.

By 4.2, $\mathbf{H} = (0, 0, 6, 250)$, $G = 15{,}008$.

Note that answers are sums of answers to Ex. 3.1(b, c).

b. $\mathbf{x} = (x, y, x^2 - y^2, 2xy)$, $\mathbf{x}_1 = (1, 0, 2x, 2y)$, $\mathbf{x}_2 = (0, 1, -2y, 2x)$,

$\mathbf{x}_{11} = (0, 0, 2, 0)$, $\mathbf{x}_{12} = (0, 0, 0, 2)$, $\mathbf{x}_{22} = (0, 0, -2, 0)$.

By 4.2, $\mathbf{H} = P \dfrac{\mathbf{0}}{\text{something}} = \mathbf{0}$. For G, we need $P\mathbf{x}_{ij}$.

Since $\mathbf{x}_1$, $\mathbf{x}_2$ are orthogonal,

$$P(\mathbf{x}_{11}) = \mathbf{x}_{11} - \frac{\mathbf{x}_{11} \cdot \mathbf{x}_1}{\mathbf{x}_1^2} \mathbf{x}_1 - \frac{\mathbf{x}_{11} \cdot \mathbf{x}_2}{\mathbf{x}_2^2} \mathbf{x}_2 = \cdots$$

$$= \frac{(-4x, 4y, 2, 0)}{1 + 4x^2 + 4y^2},$$

$$P(\mathbf{x}_{12}) = \frac{(-4y, -4x, 0, 2)}{1 + 4x^2 + 4y^2}, \; P(\mathbf{x}_{22}) = \frac{(4x, -4y, -2, 0)}{1 + 4x^2 + 4y^2}.$$

$$G = -8(1 + 4x^2 + 4y^2)^{-3} = -8(1 + 4|z|^2)^{-3}.$$

4.2. $\mathbf{x} = (z, f(z))$, with $z = u_1 + iu_2$.

$\mathbf{x}_1 = (1, f'(z))$, $\mathbf{x}_2 = (i, if'(z))$, $\mathbf{x}_1 \cdot \mathbf{x}_2 = 0$ $(\mathbf{x}_2 = i\mathbf{x}_1)$,

$\mathbf{x}_{11} = (0, f''(z))$, $\mathbf{x}_{12} = (0, if''(z))$, $\mathbf{x}_{22} = (0, -f''(z))$.

$$\mathbf{H} = \frac{P(\mathbf{0})}{\text{something}} = \mathbf{0}.$$ For G, we need $P\mathbf{x}_{ij}$.

Since $\mathbf{x}_1$, $\mathbf{x}_2$ span a complex subspace, P just projects onto the orthogonal complex subspace spanned by $(-\overline{f'(z)}, 1) \in \mathbf{C}^2$.

$$P(\mathbf{x}_{11}) = -P(\mathbf{x}_{22}) = \frac{(-\overline{f'}f'', f'')}{(1 + |f'|^2)},$$

$$P(\mathbf{x}_{12}) = \frac{(-i\overline{f'}f'', if'')}{1 + |f'|^2}.$$

$$G = -2|f''|^2(1 + |f'|^2)^{-3}.$$

4.3. Calculating with formula 4.2(1) yields

$$\mathbf{H} = P\mathbf{v}(1 + |f_x|^2 + |f_y|^2)^{-1},$$

where

$$\mathbf{v} = (1 + |f_y|^2)f_{xx} - 2(f_x \cdot f_y)f_{xy} + (1 + |f_x|^2)f_{yy}$$

$$\in \mathbf{R}^{n-2} \subset \mathbf{R}^2 \times \mathbf{R}^{n-2}.$$

Clearly if $\mathbf{v} = \mathbf{0}$, then $\mathbf{H} = \mathbf{0}$ and the surface is minimal. On the other hand, if $\mathbf{H} = \mathbf{0}$, $\mathbf{v} \in \ker P \cap \mathbf{R}^{n-2} = \{\mathbf{0}\}$.

5.1. a. $\mathrm{II} = \left[\dfrac{\partial^2 y}{\partial x^2}\right]_0 = \begin{bmatrix} (2,6) & (2,0) & (0,0) \\ (2,0) & (2,2) & (0,2) \\ (0,0) & (0,2) & (10,2) \end{bmatrix}.$

b. $K(e_1 \wedge e_2) = (2,6) \cdot (2,2) - (2,0) \cdot (2,0) = 12.$

c. One orthonormal basis is $v = (1, -1, 0)/\sqrt{2}$, $w = (0, 0, 1)$.

$$K = \mathrm{II}(v, v) \cdot \mathrm{II}(w, w) - \mathrm{II}(v, w) \cdot \mathrm{II}(v, w)$$
$$= (0, 4) \cdot (10, 2) - (0, -\sqrt{2}) \cdot (0, -\sqrt{2}) = 6.$$

One orthonormal basis for $\{x_1 + x_2 + x_3 = 0\}$ is $v = \dfrac{(1, -1, 0)}{\sqrt{2}},$

$$w = \frac{(1, 1, -2)}{\sqrt{6}}. \qquad K = 0.$$

d. $R_{1212} = 12$, $R_{1213} = 12$, $R_{1223} = 0$, $R_{1313} = 32$, $R_{1323} = 20$, $R_{2323} = 20$. Rest by symmetries.

Redoing b, c, and e, we have the following:

b. $R_{1212} = 12$.

c. $v_1 = -v_2 = 1/\sqrt{2}$, $w_3 = 1$, rest 0.

$$K = \tfrac{1}{2}R_{1313} - \tfrac{1}{2}R_{1323} - \tfrac{1}{2}R_{2313} + \tfrac{1}{2}R_{2323} = 6.$$

$v_1 = -v_2 = 1/\sqrt{2}$, $v_3 = 0$, $w_1 = w_2 = 1/\sqrt{6}$, $w_3 = -2/\sqrt{6}$.

$$K = \tfrac{1}{12}R_{1212} - \tfrac{1}{6}R_{1213} - \tfrac{1}{6}R_{1312} + \tfrac{1}{3}R_{1313} - \tfrac{1}{12}R_{1221} + \tfrac{1}{6}R_{1321}$$
$$- \tfrac{1}{3}R_{1323} - \tfrac{1}{12}R_{2112} + \tfrac{1}{6}R_{2113} - \tfrac{1}{3}R_{2313} + \tfrac{1}{12}R_{2121} + \tfrac{1}{3}R_{2323}$$
$$= 1 - 2 - 2 + 32/3 + 1 - 2 - 20/3 + 1 - 2 - 20/3 + 1 + 20/3$$
$$= 0.$$

e. $\mathrm{Ric} = \begin{bmatrix} 44 & 20 & 0 \\ 20 & 32 & 12 \\ 0 & 12 & 52 \end{bmatrix}$, $R = 44 + 32 + 52 = 128$.

5.2. $Df = \begin{bmatrix} \dfrac{\partial f}{\partial x} & \dfrac{\partial f}{\partial y} & \dfrac{\partial f}{\partial z} \end{bmatrix} = \begin{bmatrix} 0 & 2y & 0 \\ 1 & 0 & 1 \\ 3x^2 & 0 & 0 \end{bmatrix} = (0, 2y, 0)\mathbf{i} + (1, 0, 1)\mathbf{j} + (3x^2, 0, 0)\mathbf{k}.$

$$Df(0, 0, 1) = \begin{bmatrix} 0 & 0 & 0 \\ 1 & 0 & 1 \\ 0 & 0 & 0 \end{bmatrix} = (0, 0, 0)\mathbf{i} + (1, 0, 1)\mathbf{j} + (0, 0, 0)\mathbf{k}.$$

Covariant derivative $= \begin{bmatrix} 0 & 0 \\ 1 & 0 \end{bmatrix} = (1, 0)\mathbf{j}.$

5.3. Let $g(x_1, \ldots, x_n) = x_n - f(x_1, \ldots, x_{n-1})$. Then the graph of f is the level set $\{g = 0\}$. By 5.0(1),

$$H = -\operatorname{div}\frac{\nabla g}{|\nabla g|} = -\operatorname{div}\frac{(-f_1, \ldots, -f_{n-1}, 1)}{\sqrt{1 + f_1^2 + \cdots + f_{n-1}^2}}$$

$$= \operatorname{div}\frac{\nabla f}{\sqrt{1 + |\nabla f|^2}} - \frac{\partial}{\partial x_n}\frac{1}{\sqrt{1 + |\nabla f|^2}} = \operatorname{div}\frac{\nabla f}{\sqrt{1 + |\nabla f|^2}}.$$

Symbol Index

Name Index

Subject Index